THE END
OF
TIME

THE END
OF
TIME

---

by ENZO TIEZZI

WITPRESS   Southampton, Boston

Enzo Tiezzi

University of Siena, Italy

Published by

WIT Press
Ashurst Lodge, Ashurst, Southampton, SO40 7AA, UK
Tel: 44 (0) 238 029 3223; Fax: 44 (0) 238 029 2853
E-Mail: witpress@witpress.com
http://www.witpress.com

For USA, Canada and Mexico

Computational Mechanics Inc
25 Bridge Street, Billerica, MA 01821, USA
Tel: 978 667 5841; Fax: 978 667 7582
E-Mail: info@compmech.com
US site: http://www.compmech.com

British Library Cataloguing-in-Publication Data

A Catalogue record for this book is available
from the British Library

ISBN: 1-85312-931-3
ISSN: 1467-9581

Library of Congress Catalog Card Number: 2002102703

CONTENTS

PREFACE

"So one thing shall never cease to arise
out of another and life is granted to none
or freehold, to all as tenants on lease."
Lucretius, *De rerum natura*, III, 970-971.

I n people from very different backgrounds, there exist today the
seeds of a new ecological culture which will call into question the
very foundations of our present way of life. A new generation,
new individual and collective behaviour, new needs and movements
with new forms of expression are emerging from a 'green nebula'
which is sweeping Europe and the world. Unlimited material growth,
blind faith in science and the boundlessness of nature have been
exposed as myths of the technological society. New values are being
sought and the links between environmental questions and production analysed. What changes in cultural coordinates are required for
the emergence of a society based not only on social justice but also on
the quality of life? Which social agents are capable of bringing about
this transition?

An ecological culture which analyses the relationships between
man and the environment (resources) defining economic and political limits, is coming into being. This new culture is not marginal or
subordinate but plays a central role, establishing the principles upon
which to base a new economy and new politics.

On the basis of my experience in scientific research and teaching
in various Italian universities, and a particularly stimulating year
spent at Washington University with Barry Commoner fifteen years
ago at the dawn of human ecology, I shall try to give some answers to
these questions. In so doing I risk being labeled ingenuous. Within
the logic of the present obsolete system there are far too many

VII

tacticians and I am convinced that if we want to build this new idea of development, we cannot and must not sacrifice our ingenuousness.

This book unites passages dedicated to scientific and socioscientific problems with passages which try to create a bridge between the scientific and humanistic fields. The 'scientific' chapters deal with points which I consider to be fundamental for a new idea of development: thermodynamics and the concept of entropy; the biological foundations of the environment and evolution; the problems of war, energy, climate, food and resources.

The story 'My father and the machines', in the Appendix of Chapter 1, is about my grandfather. It was written by my mother. The story is given at face value, without modification or comment. It is a period snapshot exposing the myth of science and technology as experienced in the candid setting of an Italian provincial town.

The tenth chapter is an attempt at a 'new alliance' between science and humanistic disciplines. I examine authors such as Fromm, Huxley, Lawrence and Orwell whose work presages a new ecological culture. Now that 1984 is behind us, how true was Orwell's forecast?

The present moment is a difficult one for man-nature relations. We shall travel together through thermodynamics and biology and then with our obligatory companions, utopia and creativity, we shall venture upon the road of attempted unification of different disciplines. Here even language becomes a problem and we risk accusations of superficiality from the specialists.

I owe much of what is written here to exchanges of ideas, some quite lively and polemical, with many friends.

At this stage I no longer know what is mine and what is copied or adapted from dozens of documents edited together, dozens of readings assimilated down to basic concepts. I am convinced that there is no such thing as originality. Within us, ideas are a sum of other ideas, and what is personal blends with what was personal to others.

The verses of Jacques Prévert which appear in the text were added out of personal vanity: like myself and Nicholas Georgescu-Roegen, Prévert was born on 4 February under the sign of Aquarius.

from the farm at Pacina, Chianti (Siena)

to my grandchildren
Maria, Carlo, Tommaso and Teresa

CHAPTER 1
THE HORSE OF SAMARRA

A soldier in ancient Basra went to his King full of fear and said: "Save me, Sovereign, I must leave this place! In the market square I met Death dressed in black and he looked at me malignantly! Lend me your horse so that I can escape to Samarra.[1] I fear for my life if I remain here."

"Give him the best horse," said the King. "Son of lightning, worthy of a King."

Later the King met Death in the city and said, "My soldier was very much afraid. He said he met you today at the market and you looked at him malignantly."

"Oh no!" replied Death. "I was only surprised. I couldn't understand what he was doing here this morning because I am expecting him tonight in Samarra. He was miles away."

Perhaps there is also a Samarra in our destiny as men. Rushing to solve immediate problems, trusting the miraculous power of new technologies, all preoccupied in the choice of the fastest horse, let us hope we are not already on the road to Samarra where we will be defeated by basic questions that we are no longer capable of solving.

Our social and economic mentality is like the search for a horse to reach Samarra, for a technology to solve today's problems without regard for whether the solution increases humanity's problems, accelerates depletion of world resources, commits us to a one-way path of damage to the biosphere which is the environment necessary for our

---

[1] The story is liberally taken from a talk by P. Bente at the Campinas Symposium and from a song by R. Vecchioni. The ancient legend refers to the city of Samarra in the "green crescent" which stretched northward in an arc from Jerusalem, and southward again to Basra in Mesopotamia. It once had the richest agriculture in the world but is now desert. The song of Vecchioni substitutes Samarkand for Samarra.

1

survival. At the end of the road of unlimited growth, Samarra may await us.

If we closely examine the whole and the interdependence of things, using the biological time scale which is much larger than that of written history, we can see that for the first time, humanity has reached a crossroads from which there may be many roads to Samarra.

The first is the road of nuclear war. Today man is capable of erasing his own species and life itself from the Earth. Should this happen, there would not be a second chance for life to arise on this planet. Life emerged in an atmosphere low in oxygen and the presence of oxygen today precludes the possibility of starting again.

The second is the road of population increase. The population of the Earth has progressively increased but so far food and energy resources (though inequitably distributed with millions dying of hunger) have been sufficient. Now we have reached six billion; every four days there are a million more of us. We can expect the population to double in one or two generations. With such a population our energy resources will not last long and it will be impossible to feed everyone.

The third is the road of great biological imbalances. For instance, climatic variations induced by human activities which may soon render the planet uninhabitable.

The fourth is the road of energy waste. The fifth is the loss of the genetic heritage, and so on. These roads are closely linked; for example food production and waste of energy are interdependent and in turn depend upon population increase; they influence climate, deleteriously affecting food production.

In examining these points in the following chapters, I do not want to paint a catastrophic picture. It is my firm conviction that we must change route as soon as possible and set about defining a new idea of development. A culture so based will be firmly founded on *biology* and *thermodynamics*, and their fundamental relationship to the economy, the society and the means of production. My conviction is based on three points: (a) the equilibrium of nature is extremely delicate and can be irreversibly upset by man: the resources of nature are not infinite; (b) the destruction of the environment and waste of natural resources is never of long term benefit either economically or socially; (c) the false prosperity of the consumer society is based on the exploitation of three classes of people: (1) the younger generations who will be left with depleted resources and a polluted environment; (2) underprivileged groups who incur health hazards, pollution etc. without enjoying the economic advantages of consumerism; (3) the third world which pays for our consumer needs with monocultures, destruction of the natural and social environment and hunger.

These three points demonstrate that purely historical and economic analysis cannot further our understanding. We are concerned with a biological time scale but today's parameters are changing more rapidly than ever in the past: natural changes which previously occurred over thousands of years may now, with the impact of new technologies, practically occur overnight. The biological tempos between generations today are much faster; the speed with which the younger generations seize and assimilate new technologies is enormous; the rate at which production methods and relationships evolve is also growing rapidly.

The problems of nature and resources concern everyone, including the third world. In mathematical terms we must extrapolate both in space and time and take the rapidly emerging younger generations into account as social agents.

It becomes complicated if we persist in recognizing only the old social classes and try to solve problems using old economic models, whether capitalist or marxist. The main point to call into question, which is the erroneous axiom on which our reasoning has so far been based, is unlimited material growth. This not only involves rethinking production relationships, but what we produce, how to produce it, where, when etc. and much that has been taken for granted, for example the synonym of well-being and increase in gross national product or industrialization. A radical process of liberation of real human potential must be activated so that society can be based on balance with nature and 'quality of life' (which is not a fashionable phrase but the real requirement of many). The values required to realise this new model of development must be scientific rather than metaphysical, but also ethical and not materialistic. The traditional political channels are too conditioned by economic processes and schemes connected with 'growth' to courageously seek new values and to understand that reality does not consist only of production and consumption, salaries and profits but that natural equilibria and the renewal of resources, the system of living organisms and its continual reproduction, are just as important. We must now face this second order of reality which has been irresponsibly underestimated.

We must therefore completely reconsider our manner of producing and what we produce. In the chapters that follow, I shall endeavour to show that this implies a new culture which cannot be the simple sum of social justice and ecological conscience. The road in this direction is probably still very long; we have little idea yet of the crucial relationships between social development and ecology. It is true that some of the laws of physics (conservation of energy, entropy) and many basic biological principles are extraneous to political culture. The consequence is that the means of production of western capitalist and of communist countries have been both based on waste of resources, destruction of the environment and disregard for future generations.

This too will be discussed in the following chapters together with the importance of two lines of thought which have not coexisted very well in the past: the ideas of Marx and Malthus. Herman Daly[2] was right when he said that limits must be placed on inequality as sustained by Marx because social justice is a condition for ecological equilibrium in all non-totalitarian societies. He added that without population control and control of man's physical manufactures, all other social reform would be cancelled by the increasing weight of absolute or Malthusian scarcity.

---

[2] H.E. Daly, *Steady state economics*, V.H. Freeman, San Francisco, 1977.

In the transition to this new society, this new model of production and development, science and technology will play a primary role. But which role? Which science? Which technology? Commoner points out that the correct use of science is not to dominate, but to cooperate with nature. Biologists tell us that the natural equilibrium is extremely delicate and complex: too delicate to withstand facile futurology, too complex to be reduced to parameters even by the most sophisticated electronic data processing methods.

A branch of contemporary opinion feeds a tendency to 'technological hope' with the manifest or otherwise aim of perpetuating the model of waste of resources and aggression to the environment. The limits of this 'technological hope' are revealed by what we might call the *paradox of complexity*. It is said that technology must have a large scientific component because today one of the fundamental parameters of the productive system is complexity. But this requires increased specialisation and consequently loss of understanding and control of reality (which is complex) both by the people and by the specialists themselves. The more we specialise the less likely we are to be able to predict the effect of technology on nature. The fragmentation of scientific disciplines should not be regarded as inevitable; to the contrary every effort should be made to recreate the conditions for the unification of science (by favouring interdisciplinary research and scientific-humanistic interaction), so that science can liberate man and acquire its true social dimension.

An obvious contradiction to the claimed liberating action of technology is given by the so-called 'threshold effects', that is changes which occur without prior warning with serious consequences for the environment and the society. One of many examples is the large scale traffic jam which could not be predicted even a few minutes earlier. Another is the famous New York black-out of November 1965 when an area of more than 100,000 sq Km in northeast USA and Canada was inexplicably left without electricity for many hours. In a sense, the more recent ecological catastrophe in the Persian Gulf was a threshold effect, the cost of which was even felt by the countries responsible

5

for the original damage. The use of nuclear energy is a dangerous potential trigger of a threshold effect. Technological complexity increases the risk of the threshold effect and makes it impossible to predict. The complexity of nature prevents the intervention of the most sophisticated technologies to avoid or repair the damage.

In his first book, *Science and survival*,[3] Commoner justly states that, like the sorcerer's apprentice, our actions are based on dangerously incomplete understanding: we are actually performing experiments on ourselves!

Science and technology cannot progress without taking some risks, but the size and frequency of possible errors have grown with the development of science and the expansion of technology. In the past, the risks taken in the name of technological progress were restricted in space and time. The new risks are long term and of planetary dimensions. For the first time in the history of man, they jeopardize the survival of the human species. The gap between technological complexity and the understanding of its effects on nature is visibly widening. The superficiality of the technologist's knowledge of biology and the globe is directly proportional to his level of specialisation and specific culture. More than ever before technology is in the hands of sorcerer's apprentices who audaciously believe themselves capable of solving the world's complex problems.

In February 1970 *Time* magazine appeared with an unusual cover page: it was the face of Barry Commoner half white with a background of resplendent nature, half black with a background of polluted environment. The title was *The emerging science of survival* and the journalist defined Commoner as a professor with a class of many millions, a scientist who refused to live in his ivory tower-laboratory. Four years earlier I spent a year with him as a post-doctoral research associate, studying the nuclear magnetic and electron spin resonance of free radicals in cancerogenous processes. He had just written *Science and survival* and founded the journal *Scientist and citizen* (later *Environment*). I remember the day of my arrival at Washingon

---

[3]   B. Commoner, *Science and survival*, Viking Press, New York, 1963.

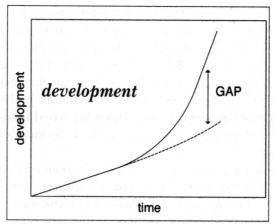

Figure 1: The graph of the sorcerer's apprentice.

University, when he took me around the Departments of Physics and Biology between which I was to divide my time. He drew a graph on the blackboard for me consisting of two curves: the graph of the sorcerer's apprentice (Figure 1). The lower curve represented the growth of our knowledge of ourselves and of biology; the upper curve which suddenly rose very steeply, accentuating the gap between the curves, was the curve of technological progress, the forward rush of the sorcerer's apprentices. Technological change is dramatically faster than natural evolution and our capacity for cultural adjustment.

Two years later in 1968, the first student protest meetings in the University of Florence proved impervious to ecological questions. My students recognised in me at least a grain of democracy and gave me the privilege of choosing the tree from which to be hanged. I replied with a story about Bert, who used the excuse of choosing a tree to tour the world, without ever succeeding in finding the right branch. Four years later *The closing circle*[4] was published by Garzanti and ecology landed on the already highly polluted Italian shores.

---

[4] B. Commoner, *The closing circle*, Alfred A. Knopf, New York, 1971.

Commoner's analysis clearly demonstrated the relationships between the ecological, energy and economic crises, showing that it was no coincidence that the three occurred at the same time. He also showed the advantage, from the point of view of the economy and employment, of using renewable energy resources, as against the negative effects in terms of unemployment, inflation and immobilization of available capital, of using non-renewable sources i.e. petroleum, coal and nuclear energy.

The scheme of Figure 2 shows the interaction between the three systems and the interdependence of the three crises. *The poverty of power*,[5] another famous book by Commoner, analyses this scheme and shows that the environmental and energy crises are the result of erroneous choices inherent in the industrial and economic systems. Energy is the key to understanding these interactions: a system based on non-renewable resources catalyses a series of chain reactions which inevitably leads to environmental damage, the depletion of resources and economic crisis.

The limits of development or rather material growth are therefore the limits of renewability (of resources, environment, energy etc.). A new model of development does not mean a return to candlepower: it is projected towards the future, not the past. It takes the biophysical equilibrium into account in dynamic, not static, terms, making its measurements in biological time and aiming at a stationary flux of energy, population and resources. The fundamental point is that growth must stop. Population, desertification, energy requirements, consumerism, pollution, climatic changes, nuclear armaments, extinct species, energy cost of food, hunger: all these must cease growing. When should they stop? Laura Conti[6] replies:

> From now on the easiest time to stop is NOW. Now is more difficult than yesterday but easier than tomorrow. Now is more difficult than a year ago but easier than in a years time. The freedom of choice decreases from day to

5   B. Commoner, *The poverty of power*, Alfred A. Knopf, New York, 1976.
6   L. Conti, *Questo pianeta*, Editori Riuniti, Rome, 1983.

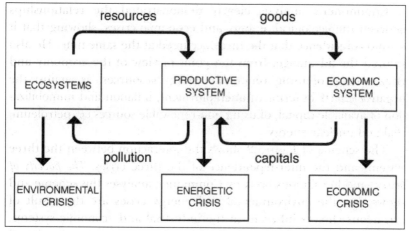

Figure 2:   The interaction between ecological, productive and economic systems.

day so there is no doubt that we must set in motion
NOW, the mechanism that will decrease our energy
requirements... The first step is cultural and consists in
negating two great myths: the myth that industrialization
is a positive value and the myth that the growth of
exchange between economies is a positive value. Import-
export figures and steel production must no longer be
quoted as an index of success: they ought to be replaced
by the river trout census or the pace of ducks.

Laura Conti's gently provocative manner suggests that the era of
the naive faith in science and technology of the youth of the twenties
is at an end. This is because science and technology today are too
powerful for a trial and error approach which could jeopardize
planetary equlibrium and the survival of future generations.

In those very years at the beginning of the century, when the
consequences of the second law of thermodynamics and the laws of
entropy defined the concept of a limit, of the inevitable degradation
of the planet, it seemed to men that (again I quote Laura Conti)
'technology itself generated more sophisticated technology and that

machines themselves generated more powerful machines. The extent of this fond delusion only began to be revealed ten years ago.'

We were woken by ecology and a global view of the Earth. We have reached the turn-off to Samarra, a decisive point in our history on this planet. The important thing is not to take the Samarra road! Turn the horse round! Which way shall we go?

APPENDIX: CHAPTER 1

The youth of the twenties and technology:
my father and machines*

My father loved machines. Any mechanical or electrical gadget that progress created evoked his admiration and sympathy. As a boy he built water or wind mills with materials that came to hand. When he had his first bicycle he at once installed a 'speed-change' of his own invention. As soon as age and money permitted, he bought a motorcycle.

In the meantime he set up a small timber mill and quickly enlarged his field of operations. From his native town of Armaiolo where his grandfather Quirino had a broom factory, the whole family moved to Rapolano. A comfortable two-storey house was built with a large terrace for my grandmother's flowers; next door was the workshop.

One problem that preoccupied my father was lighting. My grandmother told me that ever since he was a boy, he had candles and spirit lamps everywhere and still complained that it was dark. When the first mention was made of gaslight, he set to work and built some 'carbide drop' acetylene gasometers. He exhibited them at a number of 'fairs' and collected a crop of medals and declarations. Soon he had orders for illuminating private villas and even the whole town of Foiano della Chiana. At home he had chandeliers of his own design in every room and there was plenty of light. It was necessary to turn on the tap and light the nozzles with a match, but he had even invented a special lighter, a stick with a flint at its end and a wheel driven by a little chain. In this way he could light even the highest chandelier without climbing on a chair.

In the kitchen there were very fast and efficient burners. For a national celebration he built a big five-pointed star: it was a complicated tangle of gas tubes leading to an infinity of nozzles. When it was

---

* by Lidia Pulselli Tiezzi

mounted on the front of the house, the effect was beautiful: we children gazed at it with wonder. At night from my bed I listened to the hiss of the lighted flames blown by the wind. Acetylene did not cause him to neglect the workshop which grew in importance and was crowned by a chimney. I loved to watch the masons at work: as the slim four-sided tower grew, tapering towards the top, I thought it would reach the clouds. The machines were driven by steam from a big boiler, fed exclusively with sawdust and offcuts of wood. Such autonomy was very satisfying to my father. At midday a loud siren announced lunchtime. My mother and grandmother instinctively blocked their ears and if there was a stranger with us, he would jump. I enjoyed it greatly and thought that the whole town, and indeed the whole area, had gained in importance with the sound of the siren which seemed to say, 'Here too, modern industry has arrived!'

I was small, but I understood that my father was intelligent and I admired him unconditionally, but on one thing we did not agree, and that was toys. I loved wooden blocks and I would build fantastic houses with tiny dolls and papier-maché animals inside, but my father always gave me mechanical toys. Once he displayed on the dining room table a tin sheep, painted pink, mounted on wheels. It could be wound up with a key, and went forward, nodding its head. It was 20 cm high and would not fit inside my houses. It was cold and hard and its wheels set it apart from the animals. To me, it was horrible!

I liked my little papier-maché pets but I also liked real animals. We had four horses in the stable at the time: there was Tella, a graceful bay mare, who was used only by my grandfather. She used to take him around to nearby customers and suppliers in the gig. It was often she who brought him back home, because my grandfather was always writing in his notebook and thinking of business. Blind to the outside world he would let go of the reins and entrust himself to the good sense of the horse. Then there was Biancone, a big docile old whitish horse. Harnessed to a plank with four wheels he used to take my grandmother and we grandchildren for rides. Reno and Moro were two strong cart-horses used to transport tree trunks from the forest to the workshop and the timber planks from the workshop to the railway station. My father was not interested in the horses except in as far as they served his work.

I was four (it was 1909) when my father bought his first automobile: it was an event not only for our family but for the whole town. Many had only ever heard about cars. It was a big round white one with a folding roof like those of carriages. It was a four-seater but seven people comfortably fitted inside. To start it, you had to go to the front and work the crank which was a nuisance and tiring. There were no side doors, but there was little likelihood of being thrown out at the speeds possible on those roads! In summer the all-enveloping dust; in winter the cold! At all times the continual jolting on terrible pointed rocks! Animals were terrified of the roaring motor and often tried the patience of the driver. When we had to overtake a cart drawn by oxen it was a disaster. The beasts took fright and stood across the road or bolted into the ditches; the peasants were even more afraid, lost control and panicked. However in only a few years, even the animal kingdom submitted to the reign of machines.

How wonderful to travel by car! In a few hours we could go to our aunts in Arcidosso. This would otherwise have taken all day: train from Rapolano to Mt. Amiata station, change at Asciano, then an interminable climb by coach up to Arcidosso. We could also go to the seaside, to Forte dei Marmi! The trip was very long but interesting. When we went through a town, women and children appeared in the doorways and even the men turned an eye. A car going past was something that didn't happen very often! In half a day we would reach the square of Forte, in front of the Caffè Roma. The whole town would crowd around us. This was 1909 or 1910 I think.

At Forte there was the fort, the Caffè Roma, the Pensione Idone, about twenty houses and large lots full of slabs and blocks of marble ready to be loaded on three-masted ships. The wharf, with its rickety uneven planks extended out into the sea on great pylons. There was almost no-one but ourselves who came to enjoy the beautiful clean sea and the expanse of immaculate sand, and no-one had a car. There was a boy who had a kind of motorboat and he and my father became friends at once. Their passion for motors brought them together. We children liked the smell of petrol. My grandmother would pull a face and my mother found it quite unpleasant. But to us it recalled inebriating drives across unknown country, counting down long rows of trees.

My grandfather died in 1917 when I was eleven. After that the horses were sold and replaced by a truck. My father chose a big American one which had four wheel drive, and instead of being in front under the bonnet, the motor lay under the driver's seat. To me it had a monstrous appearance, like a great vulgar bulldog with a flat face. Since it was very difficult to obtain spare parts, my father bought another the same, and if something went wrong he and Dide, the driver-houseman, would fix it. There were hardly any garages and the mechanics had little experience; petrol stations were a dream of the future. I remember being woken more than once around eleven or midnight, by people beating on the front door. It was always a friend who had run out of petrol. My father was usually still up at that hour, and was happy to go down to the garage and help the stranded motorist. He was provident and always kept a good stockpile of petrol in drums.

All this time his workshop grew: a whole new building was added for the saws. Trucks loaded with trunks to be sawed into planks arrived every day in the timber yard. The planks went to the next shop where other mechanical saws, lathes and planes made the articles that my father supplied to various industries: wheelbarrows, handles of many types, cogs and other things. He had about 50 men. My father was competent, he taught them well and was not afraid to dirty his hands using the machines and working with his men. Many of the machines he had actually conceived, designed and built himself with the help of his blacksmiths. He was always trying to make improvements in the field of machines and inventions.

Already electricity had taken over in many areas and my father was fascinated. Gas pipes disappeared from our house and were replaced by electric wires. He even produced the electricity himself: using a boiler in the workshop, he charged a battery of accumulators (these large glass containers occupied a whole room) which provided the whole house with light and energy.

There began to be talk about radio transmission. My father launched into the field with enthusiasm. He bought many texts on the subject and contacted Milanese firms and amateurs in all parts of Italy. The first time we heard the voice of a radio was in Milan at the home of an engineer friend. We managed to tune into Radio Toulouse: there were various noises, crackling, murmuring and every so often a bit of music faded in and out or the voice of the announcer with a typical Midi accent. Not much, to be sure, but progress! Soon my father began to build receivers which worked well. When Radio Rome began transmission we could regularly tune in to interesting programmes and in the evening there was always a group of friends from the town who came to listen to the novelty.

My father invented a new type of condenser and took out a patent. We often had guests who were technicians or amateurs from Milan or Turin and the discussions were endless. 'After the telegraph and the telephone came the wireless and the radio,' he would say, his eyes shining, 'and soon we will have the transmission of pictures and who knows what else!' He spoke to me of Edison and Marconi: 'They may not be pure scientists, but they know how to apply scientific principles and this is important for humanity.'

He had never been in an aeroplane, but was a great admirer of aviation. He followed with interest the enterprises of Lindbergh and De Pinedo and Balbo's Atlantic crossing made him forget his daily invectives against Mussolini.

Then he fell ill: he never saw television, he didn't live to see the moon landing, it was too late to get involved in nuclear problems. He died convinced that scientific progress was everything for civilization.

I always admired and respected him but once made a comment that amazed and hurt him. He was opening a parcel of tools sent to him from Milan. 'See,' he said. 'I supplied this company with articles which they needed to make these tools and now the tools are useful to me.' 'Then it is a closed circle!' I said. 'What's the point?'

CHAPTER 2
SOMETIMES DISORDER
DEGENERATES INTO ORDER

Die Energie der Welt ist Konstant (the energy in the world is constant; R.Clausius 1865). The first law of thermodynamics says that the total energy existing in the universe in its various forms is invariant; it can only be transformed from one form to another but the total of the various forms remains constant: this first law is also known as the conservation of energy.

Julius Robert Mayer was the philosophical father of the first law. He sailed from Rotterdam in February 1840 on the *Java* as ship's doctor. After a long voyage the ship berthed at Surabaya. Here some of the sailors required medical assistance and Mayer noticed the bright red colour of their blood. A local doctor told him that the colour was typical of blood in the tropics because the quantity of oxygen necessary to maintain the body temperature was less than that required in a colder climate. Mayer began to think: evidently with the same amount of food the body could produce a variable amount of heat, and the body also performed work. If a given quantity of food gave a fixed quantity of energy, the body could use it for heat and/or work; heat and work are interchangeable quantities of the same type but their sum is constant. The beautiful, precise experiments of the Englishman J.P. Joule, performed in his father's beer factory, confirmed and defined the first law.

Given that a particle contains energy proportional to its mass (Einstein's law: $E = mc^2$ where E is energy, m mass and c^2 the proportionality constant, the square of the speed of light) and that the conversion of mass to energy can be measured in nuclear reactions, the first law can also be regarded as the conservation of mass and energy.

The first law says that a machine cannot create energy. Lazare

Carnot was involved with machines and their energy yield between one reorganisation and another of the armies of the Republic. His son, Sadi Carnot, wisely followed in his father's footsteps, studying the energy output of thermal machines and discovered the second law of thermodynamics. This was in apparent conflict with the first law. The second law says that energy cannot be freely transformed from one form to another and that thermal energy (heat) can flow freely from a hot to a cold place but never in the opposite direction. The conversion of heat into work is impossible without a temperature difference. The second law of thermodynamics says that a machine cannot transfer heat from a cold body to a hot one without performing work. Whenever work is produced from heat, heat also passes from a hot body to a colder one. Our everyday experience with devices (from motors to electric razors) shows us that work is inevitably accompanied by heating which was not the intention of the machine. There is a tendency in the universe towards the 'heat form' of energy. Heat is a 'degraded' form of energy because it cannot be totally converted back into work. Only some of the heat can be transformed into work; we cannot freely recover heat from a cold body. For example the ocean is an immense store of heat, it contains an enormous quantity of energy, but we cannot use it gratis. Although it contains much more heat than our bodies, we cannot warm our hands by it because the ocean is a colder source than our hands and heat cannot pass spontaneously from a cold to a warmer body.

At this point we could open a parenthesis on what we might call 'Maxwell's demon revisited'. In 1871 J.C. Maxwell proposed a paradox which embarrassed physicists for a long time. He imagined a system with gas in two containers A and B at the same temperature, separated by a wall. There was a small aperture in the wall guarded by a demon who separated fast from slow moving molecules (i.e. hot from cold molecules as temperature is a measure of the movement of molecules) putting the first into A and the second into B. In the end there would be a temperature difference in contradiction of the second law of thermodynamics. N. Georgescu-Roegen[1] observes that

18

we now consider Maxwell's demon to be exorcised; like any other living creature, he must use more energy than he creates by separating hot and cold molecules. He adds that many theories on the unlimited renewability of resources imply a demon having miraculous faculties behind the scenes.

The first law of thermodynamics is concerned only with the general energy balance, stating that this can neither be created nor destroyed. The second law is concerned with the use of energy, its availability for work and its tendency in nature to degenerate into non-utilizable forms. The thing that diminishes in the world is not the amount of energy, but its capacity to perform work. From this point of view Einstein is right to regard the second law as the fundamental law of science; Commoner is right to call it our most profound scientific intuition on the functioning of nature; and C.P. Snow is right in saying that to ignore the meaning of the second law of thermodynamics is like admitting to never having read a single work of Shakespeare.[2]

WHAT IS ENTROPY?

The spontaneous tendency of energy to degrade and be dissipated in the environment is evident in the phenomena of everyday life. A ball bouncing tends to smaller and smaller bounces and dissipation of heat. A jug that falls to the ground breaks (dissipation) into many pieces and the inverse process which could be seen running a film of the fall backwards, never happens in nature. Perfume leaves a bottle and dissipates into the room; we never see an empty bottle spontaneously fill. There is thus a tendency to the heat form and dissipation. The thermodynamic function known as *entropy* (S) is a measure of the degree of energy dissipation. Transformations tend to occur

---

[1] N. Georgescu-Roegen, was a scientist of Rumanian origin who taught economics in the US. He is known for his application of the principles of thermodynamics to economics. He is the author of *Energy and economic myths*, Elmsford, Pergamon Press, New York, 1976.

[2] C.P. Snow, *The two cultures*, Cambridge University Press, Cambridge, 1964.

spontaneously in the direction of increasing entropy or maximum dissipation. The idea of the passage of time, of the direction of the transformation, is inherent to the concept of entropy. The term was coined by Clausius from $\tau\rho o\pi\eta$ (transformation) and $\epsilon\nu\tau\rho o\pi\eta$ (evolution, mutation or even confusion).

With the concept of entropy Clausius reworded the second law of thermodynamics in a wider and more universal framework: Die Entropie der Welt strebt einem Maximum zu (The entropy of the world tends towards a maximum; R. Clausius 1865). Maximum entropy, which corresponds to the state of equilibrium of a system, is a state in which the energy is completely degraded and can no longer produce work.

Entropy is therefore a concept that shows us the direction of events. Commoner notes that sand castles (order) do not appear spontaneously but can only disappear (disorder); a wooden hut in time becomes a pile of beams and boards: the inverse process does not occur. The direction is thus from order to disorder and entropy indicates this inexorable process, the process which has the maximum probability of occurring. In this way the concepts of disorder and probability are linked in the concept of entropy. *Entropy is in fact a measure of disorder and probability.* In order to understand this better, it is useful to describe a model experiment: the mixing of gases.

Suppose we have two gases, one red and one yellow, in two containers separated by a wall. If I remove the wall I see that the two gases mix until there is a uniform distribution: an orange mixture. If they were originally mixed I would not expect to see them spontaneously separate into red and yellow. The 'orange' state is that of maximum disorder, the situation of greatest entropy because it was reached spontaneously from a situation of initial order. Entropy is a measure of the degree of disorder of the system. The disordered state occurred because it had the highest statistical probability. The probability of there being 13 hearts in a hand of bridge is one in 635,013,559,600. In other words, such a hand is almost impossible; a mixed hand, with a few cards of each suit, is the most probable. The law of increasing entropy is therefore a law of probability, of statistical tendency towards disorder. The most likely state is realised, namely the state of greatest entropy or disorder. When the gases mix, the most probable phenomenon occurs: degeneration into disorder. Nobel Prize winner for physics, Richard Feynman[3] comments that irreversibility is caused by the general accidents of life. It isn't against the laws of physics that the molecules rebound so as to separate; it is simply improbable and wouldn't happen in a million years. Things are irreversible only in the sense that going in one direction is probable whereas going in the other, while it is possible and in agreement with the laws of physics, would never happen in a million years.

The principle of increasing entropy is now clearer: high entropy states are favoured because they are more probable, and this fact can be expressed by a particular relation as shown by Boltzmann,[4] in which logarithmic dependence makes the probability of zero entropy equal to one. But the 'third law' of thermodynamics as framed by W. Nernst says that the entropy of any system at the temperature absolute zero, can always be taken as zero. It follows that 'a system at zero absolute can only have a single dynamic state, namely that of minimum energy compatible with the crystalline structure or state of aggregation of the system' (Enrico Fermi).

Clearly the 'third law' is a limit law which only attributes zero entropy to ideally crystalline substances at absolute zero. It follows that everything which exists has positive entropy.

A fourth law of thermodynamics was recently introduced by Georgescu-Roegen to broaden the theory of entropy. This law states for matter what the second law says for energy. It says that an isolated system tends towards chaos, towards a 'non-availability' of energy-matter. Georgescu-Roegen observes that rubber molecules abraded from tyres, phosphorus molecules dissipated by fertilizers, copper molecules consumed from the surface of coins and so forth, cannot effectively be reutilised. Dispersed matter cannot be recycled.

---

[3] R. Feynman, *Elementary particles and the laws of physics*, Cambridge University Press, Cambridge, 1991.

The main point is still the idea of entropy. A. Eddington writes that from the point of view of the philosophy of science, the concept of entropy must be regarded as the greatest contribution of the nineteenth century to scientific thought. The universality of the law of entropy increase was stressed by Clausius in the sense that energy is degraded from one end of the universe to the other and that it becomes less and less available in time, until 'Wärmetode', or the 'thermal death' of the universe.

Evolution towards the thermal death of the universe is the subject of much discussion. In order to extend the theory from the planetary to the cosmic context it is necessary to introduce unknown effects such as gravitation. Current astrophysics suggests an expanding universe which originated in a great primordial explosion (big bang) from a low entropy state, but the limits of theoretical thermodynamic models do not allow confirmation or provide evidence.

The study of entropy continues: this fundamental concept has been applied to linguistics, the codification of language and to music and information theory. Thermodynamics has taught us two fascinating lessons: that energy cannot be created or destroyed but is conserved, and that entropy is always increasing, striking the hours of the cosmic clock and reminding us that both for man and for energy-matter, time exists and the future is distinct from the past by virtue of a higher value of S.

THE RULE OF IRREVERSIBILITY AND UNCERTAINTY

Thermodynamics reigns throughout science from mechanics to nature but the natural biological laws of evolution seem to contradict it. Biological systems apparently violate the second law, they show

---

[4] According to the formula of Boltzmann, $S=k \log \pi$, where k is Boltzmann's constant ($k = R/N$, where R is the universal gas constant and N Avogadro's number) and π the state probability. (Strictly π is the number of dynamic states, or discrete quantum states corresponding to the thermodynamic state in consideration.)

extremely ordered structures and evolve in a direction of increasing order or less entropy. The contradiction is really only one of appearance. The entropy balance must be total and must include both biological organisms and the environment with which the organism continually exchanges energy and matter. Thus biological organisms develop and live by virtue of the increased entropy which their metabolism provokes in the surrounding environment. The total change in entropy is positive: the entropy of the universe increases and the second law is not violated.

If bacteria are allowed to reproduce in glucose solution, part of the sugar can be seen to cause a decrease in entropy, being transformed into cell components. The rest is transformed into carbon dioxide and water, leading to an overall increase in entropy.

It is necessary to distinguish between *isolated systems* (which cannot exchange energy or matter with the outside world), *closed systems* (which can exchange energy but not matter, e.g. our planet) and *open systems* (which can exchange both energy and matter). Cities and biological organisms are examples of open systems. For such systems we must sum the negative entropy produced inside the system with the positive entropy created in the environment. We then see that if 'sometimes disorder degenerates into order' this is only a facade, the appearance of order at the price of even greater disorder in the surrounding environment. Living systems therefore need a continuous flux of negative entropy from the universe, to which they return an even larger amount of positive entropy. I. Prigogine called these open systems 'dissipative structures'. The flow of energy causes fluctuations in the dissipative structure which reorganises tending to a higher level of complexity. As such it requires an even higher energy input and is even more vulnerable to fluctuations. It reorganises again in continuous biological evolution towards complexity and higher energy needs. All this occurs despite the environment, or rather in favour of its entropy. There is no need to invoke chance (as did Jacques Monod[5]) to explain biological

---

5   J. Monod, *Il caso e la necessità*, Italian translation, Mondadori, Milan, 1970.

evolution.[6] The thermodynamics of open systems, dissipative structures, says that the most likely event always occurs. The question is whether or not the action of man must favour the development of dissipative structures, and hence biological evolution and the dramatic increase in entropy in the environment. It is not necessary to question evolution or thermodynamics. This issue will be discussed in the next chapter. Now it is enough to say that if we increase the order and energy flow in certain living systems, even greater disorder is produced in the environment, promoting conditions which will make it impossible for man to survive.

The role of entropy in biological evolution is obviously fundamental. Entropy brings into biology the concept of a direction in time, namely that time prefers to flow from past to future, from lesser to greater entropy. Time and complexity become the protagonists of our era, imposing different frameworks for natural, economic and social phenomena. Nature shows us that her processes are irreversible. *Irreversibility and uncertainty are the rule*, says Prigogine.[7]

Uncertainty had already been introduced into physics by Werner Heisenberg in 1927 (he was only 27; a few years later he received the Nobel Prize). It was an ingenious intuition which stimulated much new work and opened new horizons.[8]

---

[6] On this point Prigogine writes in *La nouvelle alliance*: " Life, regarded as the result of <improbable> initial conditions, is in this sense compatible with the laws of physics (the initial conditions are arbitrary), but does not follow from the laws of physics (which do not set the initial conditions). This is the view of Monod, for example, in his book (5). Furthermore life, from this point of view, looks like a continual struggle by an army of Maxwell devils against the laws of physics, to maintain the highly improbable conditions which allow it to exist. Our point of view is completely different in that vital processes, far from being outside nature, follow the laws of physics, though in specific non-linear interactions and in conditions far from equilibrium. These aspects can in fact provide the flow of energy and material necessary to build and maintain functional and structural order".

[7] I. Prigogine and I. Stengers, *La nouvelle alliance. Métamorphose de la science*, Gallimard, Paris, 1979.

[8] W. Heisenberg, *Tradition in der wissenschaft*, R.Piper & Co., Verlag, Munchen, 1977.

Heisenberg discovered that the precise observation of the elementary particles of which matter consists is such that they are affected by the means of observation. It is impossible to define reality exactly: the velocity or position of a particle can be determined but not both at the same time. We can know exactly where the electron is but not how fast it is moving; or vice versa we can know its velocity but not its position.

Heisenberg described this situation with his famous inequality which he called the *uncertainty principle*:

$$\Delta p \cdot \Delta q \geq h$$

where Δp is the uncertainty in position, Δq the uncertainty in momentum and h Planck's constant. A crude example of this principle could be as follows. Suppose we want to determine the trajectory and velocity of a projectile. If we use many screens perforated by the projectile to indicate its trajectory in space, its velocity will be affected. With an infinite number of screens the error in the trajectory will be zero but the error in the velocity enormous. Vice versa with no screens there will be no error in velocity but we will know nothing about the trajectory.[9]

---

[9] The uncertainty principle can also be expressed in terms of time and energy: $\Delta E \cdot \Delta t \geq h$, where ΔE is the uncertainty in energy and Δt the life of a given state. In a system which varies with time it is impossible to determine exactly the energy of its quantum states, since if the state of the system varies at a velocity $1/\Delta t$, its energy levels are determinable with an error of the order of ΔE. Only in the hypothetical case of a state with an infinite life could its energy be determined exactly. The uncertainty relation can be used for a game to do with entropy, waste of resources and foolish use of non-renewable energy (see Chapter 5). Let us suppose that ΔE is the energy available for man and Δt the time of transition to a new model of development based on minimum entropy, the elimination of waste and renewable energy: the shorter this time, the greater the energy available for future generations. With another game of extrapolation we can also apply the principle of uncertainty to knowledge: the more we specialise and deepen our knowledge in a certain field, the greater likelihood of error because the correlations between different systems and sciences are lacking. Knowledge of reality is knowledge of complexity; conversely, the broader our knowledge of many different fields, the greater the likelihood of error because of superficiality.

The uncertainty principle is formulated on a new basis that could be termed 'theory over empiricism'. It states that any experiment aimed at knowing reality, modifies that reality. Any experimental apparatus disturbs the phenomenon observed and the principle defines the inevitable link between the experiment and its interpretation. In agreement with Einstein (with whom he in fact had many arguments) Heisenberg claimed that without a theory it is not even possible to plan an experiment and that the history of science was mainly a history of ideas.

'Theory over empiricism' also seemed to influence Francis Crick, Nobel Prize for biology, who with Jim Watson discovered the double helix structure of DNA[10] (probably the most significant discovery in biology since Darwin): 'Jim (Watson)', says Crick, 'was always clumsy with his hands. You only had to watch him peel an orange'. He also admitted, in an interview in *Nature* (26 April 1974, p. 766) on the twentieth anniversary of their discovery, the truth of Barry Commoner's criticism that 'physicists oversimplify biology'.

Without detracting from Heisenberg's great intuition that any experimental fact changes the reality which it aims to study, we may well reflect upon the dangers of reductionism and the inadequacy of mathematical models in a science as complex as biology.

THE CONTRASTING TIME SCALES OF TECHNOLOGY AND ENTROPY, ECONOMICS AND BIOLOGY, HISTORY AND 'OTHER STORIES'

Entropy means waste of resource and pollution, energy crises and destruction of the environment. The fundamental Malthusian dilemma of today is related to entropy increase. Georgescu-Roegen and J. Rifkin[11] have brought these concepts into the field of economics.

E. Schumacher,[12] Commoner and the European green movements have converted these theories into political issues.

---

[10] J.D. Watson, *The double helix. Being a personal account of the discovery of the structure of DNA*, McMillan Publishing Company, New York, 1980.

[11] J. Rifkin, *Entropy – a new world view*, Viking Press, New York, 1980.

[12] E. Schumacher, *Small is beautiful: economics as if people mattered*, Harper and Row, New York, 1973.

Dominant economic theory, based as it is on mechanistic principles, remains ignorant of the law of entropy and the role of the time variable. The classical dynamic concept of time and its reversibility, has nothing to do with reality and nature. Time is not without its preferred directions (it is not isotropic) as is space. Time has a direction. Thermodynamics introduces 'knowledge of the unidirectional flow of time', traces the limit between past reality and future uncertainty, indicates the orientation of time in natural processes. Prigogine speaks of three levels of describing time: the mechanical level described by classical and quantum mechanics, which links time to movement; the thermodynamic level which implies causality and introduces the notion of irreversibility; the level of 'dissipative structures' which introduces the notion of evolution and history.

Not only is economics ignorant of these concepts, but it introduces another which can be summed up as 'time is money'. Progress is measured by speed of production and it has even been suggested that the faster we use up nature's resources, the greater the advance of progress. In other words, the faster we transform nature, the more time we save. This technological or economic concept of time is exactly the opposite to 'entropic time'. Nature obeys different laws to economics, it works in 'entropic time': the faster we consume natural resources and the energy available in the world, the less time is left for our survival. 'Technological time' is inversely proportional to 'entropic time', 'economic time' is inversely proportional to 'biological time'.

Our limited resources and the limited resistence of our planet and its atmosphere clearly indicate that the more we accelerate the energy and matter flow through our Earth system, the shorter is the life span of our species. An organism which consumes faster than the environ-

---

[13] While the Italian original of this book was in press, a fable by Federico Butera came out in *Papir*, the Sicilian ecology journal (**3**, February-March 1984). In 'Will the anthropoman be green?' the hope is expressed that the present society, called anthroposaur, develops a nervous system and a brain and becomes an anthropoman, green in colour and perfectly integrated in the biosphere. My friend Baccio Baccetti is more pessimistic. He says that man is too big to survive, like the dinosaurs. Some small insect or water beetle will develop a miniaturized brain and will have more chance of surviving.

ment produces cannot survive, it has chosen a dead branch of the evolutionary tree; it has chosen the road taken by the dinosaurs.[13]

Money time and clock time are not the scales on which a correct relationship with nature can be established. Paradoxically the clock which is a symbol of order, strikes the hours of disorder; frenetic consumerism and growth of production bring closer the hour of global disorder. The natural order has other rhythms, another time scale. Man cannot stop time but he can slow down the process of entropy and evolution which will favour a transiton to a state of minimum entropy production and, in the long run, favour the future of our species.

Another interesting link between entropy and time concerns the environmental impact of technology. We shall look at this problem in the next chapter, and here merely underline the asymmetry of the ecological and historical time scales: millions of years for the evolution of life on the Earth with extremely slow ecological changes and historical knowledge of only the last brief period (a few thousand years); in contrast to this, the rapid ecological changes induced by technology in very short historical time. The lesson of history does not suffice to predict the future biological equilibrium including that of our species: a hundred years of history might help us plan a single day of ecological history.

Our times are thus characterised by uncertainty. Information plays a fundamental role: it becomes a primary resource like energy and materials. Entropy is a symbol of our times by virtue of its parallels with information (information as order or negative entropy) and of the continuous uncertain evolution of entropy-associated concepts which clearly show that every extension to a theory involves a change in the theory. We need a spirit of adventure to open the door and face 'other stories', issues of complexity, scarcity and uncertainty: not necessarily ugly issues.

More than a hundred years have passed since Max Planck, completely alone in the Vienna Meeting, underlined the singularity of the heat form of energy; it is more than a hundred years since Sadi Carnot, at the age of 28 years, opened the way to the concept of entropy. The dominant socioeconomic theories continue to ignore the existence of entropy: they have condemned it to a hundred years solitude, like the family of Buendiá in the novel of Garciá Márquez.

CHAPTER 3

FROM CLASS CONSCIOUSNESS TO CONSCIOUSNESS OF SPECIES: THE LEFT NEEDS BIOLOGY

We saw in Chapter 1 that for the first time in human history, crises which could involve the entire planet are foreshadowed. The population explosion (treated in detail in the appendix to this chapter), the possibility of permanent damage to the Earth's atmosphere and climate (see Chapter 7), the risk of nuclear war and the depletion of energy resources are the four most dramatic aspects of a global (environmental, energy, economic) crisis which would upset biological equilibrium. Such a crisis would be the logical consequence of the indiscriminate use, both from the biological and physical-thermodynamic points of view, of the world's resources (wrongly believed to be inexhaustible), of nature (wrongly believed to be an indefinitely self-redeeming system) and of man (wrongly believed capable of withstanding chemical and psychological assault or at least of dominating planetary scale upheavals by virtue of his knowledge and technology).

During a meeting of the editorial board of the magazine *Sapere*, a well-known Italian (anti-nuclear!) physicist played down the thermal effects of the discharge of hot water from nuclear power stations into the Tyrrhenian Sea. He argued that there was a large temperature difference between summer and winter anyway, so what difference would the few degrees caused by the hot water make? 'If you are worried about the fishermen, the solution is easy: just introduce some Red Sea fish into the Tyrrhenian and the industry can rest assured,' he added. He had no idea that biological cycles (reproduction, spawning etc.) depend on precisely defined temperatures in each season and that behind the fish of the Tyrrhenian, there are biotic and abiotic conditions which ensure the equilibrium of that particular ecosystem, and thus the survival of the fish themselves.

Instead of fossilizing in a useless attitude of faith in the power of technologies which do not take the laws of physics (entropy) and biology into account, we need to remind ourselves that man is responsible for setting in motion appalling processes which threaten his very existence; and that man can and must remedy the situation and find a starting point for future action in a scientifically and socially based relationship with nature.

In Chapter 2 we saw that all human actions are governed by an inexorable law known as the second principle of thermodynamics or the law of entropy, which states that all energy passes from utilisable to non-utilisable forms and that all human activities (particularly those which create order and organisation) inevitably lead to disorder, crisis, pollution and environmental destruction. The quality of our life and the fate of the planet depend on the appropriate application of this law. The industrial revolution has accelerated the process of destruction. Man can either accelerate the process (for profit, consumerism, power) and finish off the planet in a few hundred years or slow the process down to natural rhythms offering man and nature millions more years of life.

The logical consequence of the first two chapters is research into the concept of biological equilibrium and how to maintain it. Geological, meteorological, ecological, oceanographic and biological studies demonstrate that the life of every organism is part of a large scale process which involves the metabolism of the whole planet. *Biological activity is a planetary property, a continuous interaction of atmospheres, oceans, plants, animals, microorganisms, molecules, electrons, energy and material, all part of the whole.* The role of each of these components is essential for the maintenance of life. Harold Morowitz[1] writes that the environment and living organisms are all inseparable parts of the unity of planetary processes. He goes on to say that the prolonged

---

[1]   H. Morowitz, 'Two points of view on "Life"', *Science*, **83**: 5, June 1983.

activity of the global biogeochemical system is more characteristic of life than the individual species that appear, prosper for a period and disappear in the course of evolution.

Unity means complexity and complexity is necessary for the life of living systems: simplification means instability, weakened defences, deterioration. The correlations between the constituents of the natural system, the diversification, the individuality, and thus its complexity, allow it to be more flexible, to adapt to environmental changes, to have a greater probability of surviving and thus evolving. Vice versa, specialisation of the natural system means poverty of internal variations and greater vulnerability.

Federico Butera[2] goes deeply into these questions, starting from Prigogine's studies of 'dissipative systems'. Ecosystems having uniformly and generously distributed resources favour the evolution of species which specialise in the maximum exploitation of certain resources: we shall henceforth refer to such ecosystems as 'systems for specialists'. By contrast, ecosystems having few and heterogeneously distributed resources favour the evolution of generic species, henceforth referred to as 'systems for generics'. Butera claims that the industrial revolution has given rise to a system in which there is an abundance of homogeneously distributed resources. Such a system favours specialists: we no longer have to labour to diversify and seek resources. One can concentrate one's effort by specialising in a single resource and in this way make maximum profits. The modern industrial system has many characteristics of the 'system for specialists' including its instability, but we shall discuss this later.

From the biological point of view, increased complexity of relationship and increased diversity of genetic information means increased stability of the ecosystem. 'Biological complexity' is thus synonymous with stability.

Man's technological capabilities today have created an artificial system with an enormous potential to modify nature. This type of

---

[2]   F. Butera, in A. Baracca and E. Tiezzi, ed., *Entropia e potere*, pp. 35-46. CLUP/CLUED, Milan, 1981.

modification leads to the destruction of biological species and genetic heritage. This in turn means diminished biological complexity, reduced diversification and adaptation to change, the uncontrolled increase of certain populations generally of poor genetic heritage, and vulnerability.

Vulnerability of the oceans and seas: chemical fertilizers washed to the sea in increasing quantities from industrialised agriculture and the phosphorus of city wastes (detergents etc.) lead to eutrophisation. This is the supernutrition of the algae and bacteria involved in the process of decomposition: the phosphorus is a rich banquet for these organisms which multiply out of all proportion depriving the fish of oxygen. The fish then die in large numbers and collect in vast putrifying heaps. Eutrophization is already occurring in the Adriatic.

Vulnerability of agriculture: speaking of Valtellina, Laura Conti[3] recounts that a few years ago, the destruction of pollinating insects caused the loss of 90 per cent of the crop. In Israel, insecticides used to exterminate locusts caused the elimination of birds of prey; the resulting invasion of rats caused a much greater loss than the locusts. Our diseased forests, polluted rivers, deserts and dead lakes tell many stories of 'silent springs' and 'broken circles'. In West Africa the monocultures (all for export) create an extra 10-15 Km^2 per year of desert. In USA, concentrated agriculture produces 200,000 ha per year more desert.

Vulnerability of the atmosphere and climate: carbon dioxide, Freon and acid rain are upsetting the natural equilibria.

These forms of vulnerability affect food production. Man's attempts to produce more food often lead to a negative overall balance. Biological equilibrium is dependent upon the renewability of resources and poses limits to the growth of population and food production. For example, more fish are caught than the seas can regenerate; the speed of forest destruction (the lung of world agriculture) and increased desertification drastically reduce food

---

[3]   L. Conti, *Questo pianeta*, Editori Riuniti, Rome, 1983.

production; increased cattle raising in response to population increase only makes an apparent contribution to food production. In actual fact (see Chapter 7) the pressure of grazing causes soil erosion and the expansion of deserts.

In the light of the above, any serious socioeconomic analysis cannot set itself apart from scientific knowledge of biological equilibria and from the underlying laws of thermodynamics and concepts of limit and renewability of resources.

Now it so happens that the dominant schools of economic thought either do not know, or omit to consider the second law of thermodynamics: 'but Brutus is an honourable man'.[4] It also happens that Marxist and liberal thought, in their different perspectives of technological progress, do not allow for the complexity of biological equilibria and the natural and inevitable limits of resources: 'but Brutus is an honourable man'. It happens that the philosopher of *Questo pianeta*[3] does not know chemistry and that the architect does not know genetics: but the philosopher and the architect are honourable men.

The 'humanistic cultures' (Marxist or capitalist) are lacking a fundamental parameter in their historical analysis, namely biological time. This makes them static and extremely limited for planning the future. Biological time is the time scale of evolution, and its units are of the order of millions of years: billions of years separate us from the origin of the Earth; hundreds of millions from the appearance of algae, bacteria, trilobites, arthropods, fish; three million years since the appearance of man. But it is in biological time that we must measure the future. The breakdown of biological equilibria is inducing planetary changes in such short times that the geological clock is accelerated. Transformations which previously took millions of years can now occur (because of the imbalance induced) in decades. The consequent variations for human and social equilibria will correspond to an acceleration of millions of years of history.

---

[4] The use of Shakespeare's famous line for Antonio's speech in *Julius Caesar* is justified by H.E. Daly's and N. Georgescu-Roegen's criticism of economic theory.

In other words the biological and historical scales have been inverted. *Biological and historical tempos follow different rhythms.* Written history is a mere flash in the biological history of the Earth. In order to repair the damage of the planetary modifications already underway, the next ten years must be comparable, from the biological point of view, to millions of years. In other words, biology must have priority over normal 'historical' requirements: classical historical analysis no longer has the past and future units of measure which can tell us what is going to happen. Billions of years with complex factors and unrepeatable evolution go into the creation of the biological patrimony of a species; in the next decades man's action will be responsible for the extinction of a living species every 15 minutes.

Three million years ago a little man about 140 cm high, walking erect, with a brain of about 650 cc and a still strongly prognathic face (which however was no longer simian) wandered the Olduvai Gorge in East Africa: three million years of slow evolution characterised by technologies which were perfectly integrated with nature and great discoveries such as fire and language, all in harmony with nature. Then suddenly, only 10,000 years ago, there was the neolithic revolution (agriculture, the domestication of animals, property) and the beginning of completely different and much more rapid social processes. Finally, the industrial revolution: the time scale now is of the order of hundreds of years, infinitely short in biological time but human technologies now have a planetary scale potential to modify nature. Time is changing units in the man-nature relationship. This evolution is taking place on a logarithmic scale; it develops in a geometric progression with exponential growth (and we shall soon see that limiting factors such as population increase develop in the same way). Biological time has a mysterious asymmetry and our period is characterised by a series of problems that become evident in accelerated time. We must quickly note and understand in depth nature's signs unless we are to be the instrument of our own extinction.

The tools we need are thermodynamics and biology. A modern scientific culture cannot do without entropy and Darwin's theory of evolution. Besides, entropy and evolution have much in common. As we have already remarked, Clausius used the Greek word '*entropy*' which means change, evolution, to describe his new function.

"Ecology equals thermodynamics" proclaimed W. Jackson Davis,[5] adding that the laws of thermodynamics reign in biology, just as in physics.

Entropy, evolution: we cannot escape these laws. The processes of entropy and evolution are one-way. Time cannot be made to run backwards but we can influence the speed of these processes, namely their time derivative. Our way of living, consuming, behaving decides the speed of the entropic process, the speed of dissipation of useful energy, the period of survival of the human species. Being convinced about the second law of thermodynamics does not mean accelerating it, just as being convinced evolutionists does not mean accelerating evolution. I shall return to this question in Chapter 4. Here I would just like to emphasize the parallel between the two concepts and the role that they can play in moulding our future. If we can understand which parameters in the entropic and evolutionary processes should be slowed, this will be a step towards an overall framework in which human behaviour can be evaluated and a step in the construction of a new society of the future.

The inter-relatedness of entropy and evolution raises some complicated issues. At first sight, a period of evolution of an ecosystem or a species coincides with a build-up of negative entropy i.e. a decrease in entropy. If we look at the overall process we can see that the single organism feeds on negative entropy, creating order within itself and disorder in the environment. The small local decrease in entropy is balanced by a much larger increase in the universe. For a man to maintain his 'ordered structure' for a year, it has been estimated that he requires 300 trout; the trout in turn eat 90,000 frogs which eat 27 million grasshoppers, which eat 1,000 tons of grass.

The final result of an evolutionary process with its creation of

---

order and 'biological complexity' is always an increase in entropy in the environment. In other words natural systems 'know' thermodynamics very well and the thermodynamic efficiency of their processes is very high; the increase in entropy is reduced to a minimum. Biological evolution however means an overall increase in entropy.

Jeremy Rifkin[6] notes that biological ultraspecialisation is one of the main factors contributing to the extinction of a species and that 'systems for specialists' are the most unstable. He goes on to analyse technological society. With a series of examples and precise reasoning he demonstrates that the specialisation of our society proceeds at the same rate as the increase in complexity and centralization. Rifkin here is referring to 'technological complexity' which is completely different from 'biological complexity'. Technological complexity, according to Rifkin, is synonymous with bureaucracy, the ultraspecialised society, loss of individuality and vulnerability. Technological society has a high entropy production. The impact of 'technological complexity' on nature leads to reduced 'biological complexity' and high risk for natural systems.

For biological and social systems, simplification and extreme coercive specialisation are to be avoided. "Diversification is an important criterion in planning, once we no longer accept the myths that industrialization is an absolute positive value and that man and nature have an unlimited capacity to redeem mistakes," writes Elisabetta Donini.[7] Loss of diversification, increasing entropy and superspecialisation also mean loss of interdisciplinary understanding and fragmentation of knowledge.

When the new society emerges it will certainly be interdisciplinary with a necessary conceptual basis in entropy and evolution. The limits of the economic, humanistic and technological doctrines will find a new frontier in the 'new epistemological alliance' between man and nature.

---

[6]   J. Rifkin, *Entropy - a new world view*, Viking Press, New York, 1980.
[7]   E. Donini, 'Il Sesso della scienza', *SE-Scienza esperienza*, **4**, June 1983.

It will be necessary to seek communication between different worlds, to overcome strong and deep cultural barriers which often preclude a vision of the whole, a seeing beyond. We will have to explain to the biologist that we are convinced Darwinists but also that a too speedy evolution could be contrary to the survival of the human species or to social justice. To the physical chemist it must be explained that no-one is questioning the laws of thermodynamics but precisely because we know that man's activities create entropy, is it sometimes necessary to apply the brakes. The engineer needs to be told that we are not against development but that so-called technological progress often opposes social and biological progress and thus human progress. To the unionist and the economist it must be explained that their view of the process of production cannot ignore thermodynamics and biology. The historian, sociologist and humanist must be told that history, human relations and society cannot remain separate from biology and ecology: it is time for the integration of these two orders of phenomena in a new interdisciplinary approach to knowledge.

Again Donini's observations on a book by Carolyn Merchant[8] are to the point. She reassesses Tommaso Campanella and Johann Valentin Andrëa "whose philosophy of science was coherent with the integrity of the natural environment and human equality" whereas "the new science of capitalism shattered the integrity of nature and man-cosmos unity". "The death of nature was the presupposition for the manipulative aggression with which capitalism set about simultaneously exploiting resources and propagating the scientific ideal of knowledge of a world rendered predictable and controllable by its new passive and inanimate character. Thus in *Atlantis* Francis Bacon sketched the ideal of a society fragmented according to the hierarchy of capitalist industry, placing the task of guaranteeing progress in the exploitation of nature in the hands of scientists and technicians."

---

[8]   C. Merchant, *The death of nature. Women, ecology and the scientific revolution*, Wildwood House, London, 1979.

[9]   I. Prigogine and I. Stengers, *La nouvelle alliance. Métamorphose de la science*, Gallimard, Paris, 1979.

In *La nouvelle alliance*[9] Prigogine and Isabelle Stengers perform a similar reinterpretation of the man-nature dichotomy. Elisabetta Donini writes: "They speak in praise of the minorities, the historical agents of innovation, claiming that biological selection perpetuates the consistency of the life system rather than realising external aims. This proposal is an indictment of the reductionist and organisational hierarchy hypotheses of biological selection". The book by Prigogine (a physical chemist who received the Nobel Prize in 1977) and Stengers was acutely reviewed by the well-known writer of undoubtedly humanistic extraction, Italo Calvino: "Ten years after the publication of *Le hasard et la nécessité*[10] Prigogine answers Monod with the announcement of a 'new alliance' between man and the universe." Calvino goes on to stress that the origin of life and evolutionary events are not improbable, as asserted by Monod, but to the contrary, are the result of the thermodynamics of irreversible processes; they are not an accident of nature but can be found on the curve of nature's most logical development. Man is no longer alien to nature, or desperate: he recovers his dignity. Prigogine's starting point is the separation which began with Newton, between the world of man and physical nature. Once the distinction between science and wisdom or truth had been made, Kant sanctioned the separation between the 'two disciplines'. On this subject, Calvino inserts an interesting revaluation of Bergson. Calvino adds, "In *Le monde*, Michel Serres, known for his interpretation of Leibnitz and Lucretius, greets the publication of *The new alliance* in Lucretian prose rich in lyrical enthusiasm and density of understanding and above all with an optimism which has not been seen for a long time."

It should not be forgotten that the evolution of complex and ordered structures, which is the logical consequence of the thermo-dynamics of 'dissipative structures' (open systems which exchange energy with their environment: living beings) and thus not improbable but the main road of biology, implies an increase in entropy (disorder, degradation) in the environment. Once again, to

---

[10] J.Monod, *Il caso e la necessità*, Italian translation, Mondadori, Milan, 1970.

agree with Prigogine does not mean to favour the evolution of complex and ordered structures and accelerate the entropic process if this conflicts with the attainment of equilibrium in human society, of social justice and the survival of the species Homo sapiens.

While right-wing objections may be countered with the reply that no equilibrium or biological stability will ever be possible without social justice, it is a lot to expect the left with its prevalently historico-economic background, to incorporate thermodynamics and biology, especially when this means swallowing the doctrines of Malthus and the limits of growth. The most dangerous attitude on the left consists in delegating decisions on questions of science and technology to 'experts', to famous scientists and engineers. The ruling political class is usually of legal-economic or humanistic-sociological background and not versed in ecology. Hence biological disciplines remain at the fringe of politics and the official mentality. At the very most, lip service is paid to the environment, but when environmental problems conflict with economic advantages and employment within the existing structure, the tendency is always to play down the gravity of the environmental issue regardless of the consequences to future generations. This is true even when the consequences are economic or affect employment. The role of scientific 'experts' as delegates is fundamental in this last phase: the politician soothes his conscience by delegating to the scientist who is usually a superspecialised technocrat impregnated in the myth of the miraculous qualities of technology. Furthermore, if the 'expert' is left-wing he obviously puts the immediate employment situation on the scales and hardly ever sustainable development, even in so far as it affects future employment.

The problem is not tackled as a whole, which means considering the economic, political, sociological, biological, environmental and thermodynamic aspects.

In Italy if you ask a good politician or man of culture to talk about the 'passero solitario' you can be sure that he knows Leopardi's famous poem. If you ask him to what bird Leopardi was referring (Leopardi had a perfect knowledge of its biological habits), in 90 per cent of cases the reply would be vague or would indicate the common sparrow (Passer domesticus), presumably left behind by his companions.

The '*passero solitario*' (*Monticola solitarius*) is actually another bird altogether, and I doubt whether today at Recanati you could find a single one. The deep azure plumage of the male makes him unmistakable; he is the size of a blackbird and one of the most beautiful of Italian birds. His song is resonant and musical; he lives alone or in pairs and is insectivorous. The only way to raise this bird today consists in feeding it raw heart, salad, larvae and dead insects. Once it was common in the country and the towers of Italian towns, but now it is becoming rare because, being insectivorous, it has been hit by the heavy use of chemicals on the land and in towns.

Pascoli too was a careful observer and connoisseur of birds and plants. Since childhood he was surprised by the "mimosa which flowered my house on summer days with its pink plumes". I knew the spring mimosa with its yellow bunches. The biologically incorrect poetic licence did not convince me: it was unlike Pascoli. This year in Romagna I finally saw mimosas with veritable pink plumes flowering in mid-August (*Acacia julibrissin*). Apart from the many references in his poems, this note by Pascoli indicates the range of his naturalistic knowledge: "In the past there were also swallows, the ones with the little rust-coloured throat, the long forked tail and the sweeter, more expressive song; but they argued with their quarrelsome sisters the white-breasted swallows and flew away."

When I was a boy, hundreds and hundreds of swallows and housemartins circled my house in Chianti, 20 Km from Siena. Three years ago three couples returned to nest and this year only a single couple: yet swallows are not hunted in this area. Being insectivorous and since insecticides reach higher concentrations in swallows than in the insects themselves, they die or become sterile.

Within the Italian left, there are the first signs of an ecological movement; but one swallow does not mean spring, as the saying goes, and the prevailing attitude looks down upon 'ecologists'. Figures like

Odum,[11] Commoner, Brown,[12] Daly, Carson[13] and Rifkin who have or have had important positions on American parliamentary commissions or among the small circle of presidential consultants, are certainly absent from the Italian cultural scene. The fact that the Reagan administration practically ignored their advice, dissolving the government ecological organisation, is another serious matter which unfortunately has deep roots in American politics. The indictments of these American biologists, who speak in interdisciplinary terms, remain.

Barry Commoner commented that the system of economics and production developed without consideration for its compatibility with the ecological system. W. Jackson Davis stated that it was necessary to distribute diminishing resources ethically according to basic human needs. Lester Brown said that man's attention would return to the three basic problems: food, energy and population. Jeremy Rifkin believes that if we continue to ignore the law of entropy and its function in defining the framework of our physical world, we do so at the risk of our own extinction.

The fundamental concepts to clarify in order to approach world problems (food, energy, population, resources) in scientifically correct terms (biology and thermodynamics) are the concepts of limit and renewability. But first we need a grounding in evolution. The protagonists of evolution are adenosine triphosphate (ATP), the energy source of all living organisms, and nucleic acids (DNA and RNA). Something is known of the mechanisms and subatomic structures of these molecules, but much is yet to be discovered.

These molecules are responsible for the birth and future of life and for biological evolution. Hundreds of laboratory experiments have demonstrated the possibility of evolution of organic ('life') molecules from inorganic material plus electricity. Millions of findings have built,

---

[11] E.P. Odum, *Fundamentals of ecology*, Saunders, Philadelphia, 1971.
[12] L. Brown, *Twenty-ninth day: accommodating human needs and numbers to the earth's resources*, W.W. Norton, New York, 1978.
[13] R. Carson, *Silent spring*, Houghton Mifflin, Boston, 1962.

brick by brick, the edifice of Darwinian evolution. For millions of years DNA has acted as planner in primitive organisms like microbes and viruses and in man, transmitting genetic characteristics and acting as a blueprint for the construction of proteins. Today even the origin of the genetic code is explained in evolutionary terms (the research of M. Eigen), in terms of natural selection of the most suitable molecules and this could be one of the major tributes of modern science to Charles Darwin at the centenary of his death. Paradoxically, a hundred years ago it was thought to pay tribute to Darwin interring him in Westminster Abbey next to Isaac Newton. Without denigrating the wonderful contribution to science of Newton, they were his ideas that asserted the separation between physics and the human world, and it was with Darwin that they were reunited.

Even today there is much misunderstanding about evolution. More than once I have met people who in good faith sustained evolution as interpreted by the pre-darwinian Lamarck, in the sense of adaptation of the individual to the environment. Darwinian evolution is natural selection as a result of the struggle for survival. "Whites tan in the sun but negroes do not become fair in the shade," writes Laura Conti.[14] In other words white skin is not an adaptation of the original black man to the nordic climate, but the survival of a few fair mutants favoured by natural selection because of their ability to synthesize Vitamin D and fix calcium salts. So natural selection lies at the basis of everything.

Darwin guessed this from the essay on population by Malthus and the simple statement that the available resources are insufficient for the survival of all beings born, i.e. there is a disproportion between births and resources. The concept of limited resources emerges in this way, and is only avoided by resorting to renewable or 'eternal' resources (sun, water power, biomass etc.).

Until a short time ago the conviction that Malthusian was a synonym for reactionary prevailed in the European left but at the same time Darwin's theory of evolution was always defended with

---

[14]   L. Conti, *Che cos'è l'ecologia?*, Mazzotta, Milan, 1977.

drawn swords. The paradox is evident: evolutionism rests on the very hypothesis of Malthus. It can only be explained by the fragmentation of disciplines and unlimited faith in the sacred texts of Marx. Very few people in the Italian left understand this problem and have examined it in depth: two exceptions are Laura Conti[3, 15] and G.B. Zorzoli.[16] Some points arising from the Malthus-Marx debate are discussed in the next chapter.

At this stage I would like to emphasise three concepts: (1) In a biological system there is always disproportion between births and availability of resources until a resource becomes a 'limiting factor' of the system and of the utilization of resources; (2) Commoner showed that the Malthusian concept needs to be extended in the direction of 'decreasing productivity of non-renewable resources': coal and oil will be mined from progressively poorer and more distant deposits until the energy necessary to mine, for example, the deepest coal will be more than the energy of the coal mined, making it impossible and useless to mine it. (3) Today modern ecological studies have unequivocally shown that in the relationship between human activity and natural phenomena the productivity of energy decreases. This concept and the relative experimental data were unknown to Malthus and Marx who approached the problem only from the point of view of productivity of the earth and of labour: there was no concept of ecosystem then nor of the overall productivity of the ecosystem itself. It is precisely the decreasing yield of energy in agriculture that limits the production of food for the increasing population.

In this way we reach the inevitable 'limits to growth' (which are not necessarily limits of development), not as the result of a political ideology, but as the logical and necessary consequence of the laws of physics and biology. The most alarming problems concern energy and food, which are intimately linked and dramatically amplified by the population increase.

---

[15] L. Conti in A.Baracca and E.Tiezzi, ed., *Entropia e potere*, pp. 25-34. CLUP/CLUED, Milan, 1981.

[16] G.B. Zorzoli, *La formica e la cicala*, Editori Riuniti, Rome, 1982.

In the face of the gravity and vastness of these problems certain positions, which can be schematically grouped in three categories, become sterile and lacking in overall vision: (a) rejection of the work of the Club of Rome and of the reports it has published since *The limits to growth* in 1972.[17] This is the position of Philippe Braillard[18] in his book *L'imposture du Club de Rome*. Whereas on one hand he points out that the reports of the Club of Rome "are never completely free of a technocratic tendency characterised by lack of concern for the political and conflictive dimension of the problems", on the other he seems unaware that politics and the future of man's knowledge cannot neglect the laws of biology and thermodynamics and that no future policy can avoid facing the objective limits of resources. Zorzoli is right in noting the unjustified self importance with which too many people dismissed the first report of the Club of Rome the minute it was published: "Such an attitude on the part of the political and economic powers is understandable because even partial acceptance of the paradigm of the report would mean the radical reformulation of too many strategic hypotheses. It is less evident why this attitude is so widespread among intellectuals, men of culture and laymen with some interest in the problem. The hostility of the left, both new and old, traditional and extraparliamentary, is astounding." (b) All ideologies which view the domination of nature and technological progress as necessary for the improvement of human life, or in another field, the habit of thinking that biological evolution means progress. The role of entropy and the limits of nature today induce us to rethink our conception of evolution, progress and construction of material things. The correct use of science is not to dominate, but to live with nature. Free trade and Marxist principles are forced to admit their limits: the limits of human capacity, of nature, of the beneficial effects of industrialization. Economists continue to believe blindly in unlimited growth and technology but nature has cycles which follow

---

[17] D.H. Meadows and D.L. Meadows, *The limits to growth*, Universe Books, New York, 1972.

[18] P. Braillard, *L'imposture du Club de Rome*, PUF, Paris, 1982.

other rules and other tempos. Economists regard specialisation as positive and as a source of efficiency but nature teaches us that specialisation means risk and menace for the stability of living systems. (c) The position of those who believe that three quarters of humanity will passively accept the welfare of a minority and also pay its price. The price is consent to pollution, exploitation, monocultures etc. which accelerate the destruction of the planet and increase the gap between rich and poor countries. Any transition to a new model of development which is not based on social justice is politically impossible.

It is time for new alliances, a new culture. A very interesting starting point is that of the 'steady state' theory. The American economist Herman Daly[19] unites Marxist and Malthusian ideas in a single theory: social justice is a precondition for ecological equilibrium in all non-totalitarian societies; birth control without reform of property rights may reduce the number of poor but will not eliminate poverty.

Thermodynamics and biology dictate that we make a transition towards a state of minimum production of entropy and conservation of resources. There is very little time to bring about this transition. The critical point at which the materials-energy of the planet are exhausted to the point of preventing further change may be quite close. Let us repeat what Laura Conti had to say: "There is no doubt that from now on the easiest time to stop is NOW. Now is more difficult than yesterday but easier than tomorrow."

To create a steady state of slow growth in the society means maintaining the energy flux at a constant low level, slowing the entropic process, favouring decentralisation and the small scale, using renewable resources. On the contrary, as Butera observes "the industrial social system organised itself for an inexhaustible flow of negative entropy as big as you please and for an environment capable

---

[19] As concerns 'sustainable development' see in Robert Costanza ed., *Ecological economics*, Columbia University Press, New York, 1991, and Herman Daly, and John B. Cobb, Jr. *For the common good*, Beacon Press, Boston, 1989.

of absorbing without degradation the indefinite growth of its entropy. It is obvious that neither of these assumptions is true." The opinion of ecologists of the stature of Odum, Daly, Georgescu-Roegen is that failure to seek a steady state may mean the irreversible decadence of humanity. It certainly means ignoring historical and biological reality which is plain for all to see. However I think it is still important to repeat and underline Rifkin's idea that without a fundamental redistribution of wealth, any attempt to reduce energy expenditure and to observe the biological limits of our planet, will only block the poor in their state of subordination to the rich forever. This will trigger a series of world conflicts, the consequences of which are analysed in the next chapter.

What social agents can bring about this transition? Whoever organises the transition firstly cannot ignore the question of social justice. This implies a left-wing approach but its choices must not be subordinate to short-sighted working class, economic or historical viewpoints; rather they must encompass the wider fields of biology and the green movements. These social agents must carry the seeds of a new culture which will be interdisciplinary and comprehensive: ecological, in a word.

Let us now cite two phrases which seem fresh from the tongue of a modern ecologist but are in fact the fruit of the profound and advanced reflections of Jean-Paul Sartre.[20] "There is less and less food for human needs and there are fewer and fewer men who produce it: the shortage is real." "We must not delete all reference to the biological origin of man." The French philosopher is referring to our common biological origin and our common purpose. He identifies the brotherhood of man with the relationships between the members of a species. On the basis of this relationship Sartre launches an appeal to help us pass this 'ghastly' moment in history: "Sub-men, do not despair!" It is an appeal founded in evolution and the idea of purpose. "We are not complete men", Sartre explains. "We are beings who struggle to establish human relations and to arrive at a definition

---

[20] B. Levy, Intervista a Jean-Paul Sartre in *La repubblica*, **86**, 14 April 1980.

of man. It is a struggle that will last a long time, but it must be defined: we are trying to live together as men, we are trying to be men. This search, which has nothing to do with humanism, can lead us to define our purpose. In other words, our aim is to constitute a body in which everyone is a man and in which the collectivity is human."

Sartre's vision convinces and fascinates me. My scientific background tells me that life molecules have a common origin, as have ATP which regulates our energy supply and DNA which governs biological evolution; the same energy, material, electrons and nuclei constitute everything that exists. Sartre's analysis however is born of humanistic culture and is deep enough to suggest a true fusion of the two cultures to produce a synthesis of great clarity and breadth.

Man has a long way to go before he can behave in this way; we are not yet sufficiently evolved to have acquired a concept of fraternity, of common species. Man's actions do not yet tend towards the common aim of survival of the species. *Class consciousness* must evolve into *consciousness of species*.

The acquisition of species consciousness is fundamental for man-nature relations: the survival of the species Homo sapiens depends on our behaviour towards limited resources, the environment, population increase, war. The new highly controversial science of sociobiology is concerned with these questions. First however I would like to outline the extremely acute and interesting observations of two Italian biologists, Pietro Omodeo and Danilo Mainardi, on this problem and its connection with evolution.

Omodeo[21] recently developed the concept of fitness, studying the interactions between population and environmental resources. Given that there are species which reduce environmental resources and others which promote them, Omodeo defines 'environmental capacity' as the capacity to supply the subsistence necessary for a population of a given size. Environmental capacity and the adaptation of a

---

[21] P. Omodeo, 'La portanza dell'ambiente' in *Ambiente, risorse, salute*, **12**, 12 February, 1983, and S.Presciuttini, Intervista a P. Omodeo, in *Testi e contesti*, **7**, May, 1982.

population chase each others' tails. Omodeo says they are like the Queen of Hearts in *Alice in Wonderland*, who is forced to run forever in order to stay in the same place. In analysing the population-resources relationship, Omodeo states that "in any model, we have to avoid identifying the best, best adapted or most efficient as the individual who has the highest production costs and consequently depletes and destroys the environment for future generations." He adds that the "reproductive cost of our species is increasing from generation to generation in a terrifying way, inducing nations to escalating highwayman policies which destroy the resources of the future. The concrete prospect arising from this is the disappearance of our species." However Omodeo also speaks of 'group selection' (already suggested by Darwin) and the evolution of a community as 'superindividual'. If a community turns out to be unbalanced it becomes extinct making room for "communities whose components, for genetic reasons, behave in a more cooperative manner". "In a competition between populations of a given species, a population which leaves the environmental resources intact for the following generations will presumably have a selective advantage." I think that Omodeo's position is not very distant from Sartre's and is concerned with 'species consciousness'.

In the book *Intervista sull'etologia*[22] Mainardi approaches the question of man and evolution in these terms: "Man is effectively a great modifier of the environment. There are also other animals which have a similar property, but to a much smaller degree. Indeed I would say that man's present lifestyle seeks to modify the environment to protect his genes rather than let his genes evolve in order to adapt to environmental changes.... This modification of the environment has gradually become enslavement of the environment, its total domestica-tion.... It is doubtful that man is up to the task. Enormously complex ecological problems have been imposed upon him, perhaps fatally."

Niels Bohr was right: man is an actor and spectator of nature. His behaviour depends on environmental conditions and these in turn

---

[22] D. Mainardi, *Intervista sull'etologia*, Laterza, Bari, 1977.

are modified by his behaviour; environmental conditions influence the genetic evolution of human behaviour and human behaviour influences nature. Laura Conti was right: the history of the world, of life and of man can be written in terms of the alternation and combination of positive and negative feedback. The relationship between man and nature is extremely intricate and complex. Edward O. Wilson,[23] the founder of sociobiology was right: we are neither eidilons nor xenidrins.

Eidilons ('experts' in Greek) behave in a way which is entirely determined by their genes. Cultural influences cannot modify their choices: they derive great pleasure from music but the music programmed in their brains is always the same. *Xenidrins* ('foreigners' in Greek) behave in a way which is exclusively determined by cultural influence: in some tribes they smile for happiness, in others they smile for pain. No behaviour is innate.

Sociobiology speaks of coevolution: genetic and cultural evolution which affect each other's in turn. Biology and culture are closely connected, genes influence forms of learning and model a certain type of culture. This in turn influences survival i.e. determines which genes are destined to survive in the next generation. Wilson says however that genes have not been able to keep up with the rapid pace of technological and cultural change. It is obvious that genetic determinism (eugenics, racism etc.) are a million miles away from the positions of Wilson and the sociobiologists. Genetic influence only means that, faced with a choice, the biological properties of the brain will favour one alternative over another, but the choice is not inevitable. Genetic evolution is associated with cultural evolution via feedback mechanisms. This means for example that different populations use different words for blue but all are sensitive to radiation of

---

[23] E.O. Wilson, *Sociobiology. New synthesis*, Belknap Press of Harvard University Press, Cambridge (USA), 1975, and E.O. Wilson, *On human nature*, Bantam Books, New York, 1979, and P. Angela, 'Una Rana sul Tazebao', Interview with Edward Wilson, in *La Repubblica*, **151**, 24 July, 1982.

440 nanometres and all give a name to the colour corresponding to this wavelength. Wilson and Lumsden propose uniting the social sciences and biology in a new Darwinized human science. In this suggestion there is nothing reactionary. On the contrary, complains Laura Conti, more than a century later the controversy of Marxists versus sociobiologists echoes those of Engels versus Darwin's theory and Marx versus Malthus. Cultural and environmental determinism are just as dogmatic and dangerous as genetic determinism. To quote Wilson, Marxism risks being sociobiology without biology.

Some of Wilson's extrapolations should obviously be taken with a dose of caution. I refer to his comments on social engineering, i.e., the possibility of human intervention to modify genes and thus people's behaviour, even if Wilson intends intensification of useful characters, like propensity for altruism and cooperation. The point that interests me most is the relationship of man with his environment and the 'sub-man' behaviour of the human species in this context. Omodeo's analysis here could be very useful and the quality leap to a social culture which does not ignore biology could be the way out of the present difficult situation. It could suggest the most appropriate behaviour for our survival, for escaping biology-culture dualism, for "eavesdropping the whispers of biology within us".

A recent essay by Sabino Acquaviva[24] proposes bypassing genetic strategy, overthrowing the theories which maintain that it is impossible to change society without deviating from the laws of nature because of genetic conditioning. After all, one of the strong points of sociobiology is that it is based on the theory of evolution "without being limited to raising the smokescreen of instinct" and that it distinguishes itself from the approach of Konrad Lorenz and the mere transposition of animal behaviour to human behaviour. David Barash,[25] too, rightly claims that nature and culture always work together and all behaviour results from the interaction of these two factors. Man, the actor and spectator, lives in this play of feedback, sometimes as an

---

[24]   S. Acquaviva, *La strategia del gene*, Laterza, Bari, 1983.
[25]   D. Barash, *Geni in famiglia*, Bompiani, Milan, 1980.

integral part of nature, sometimes as external analyser and while he analyses, nature changes and so on. The criticism that sociobiology deprives us of free will and human dignity is superficial. Sociobiology certainly poses limits but these are an integral part of the natural world. Even our physical world is full of universal limits: we cannot go beyond the three dimensions of our physical being, we cannot exceed the speed of light, we cannot go below 274 degrees Centigrade, we cannot create zero pressure in any container, we cannot create zero entropy. Barasch adds that these limits do not preclude, but are probably indispensable for art and creativity. He goes on to say that like the Cheshire cat, sociobiology cannot tell us which road to take. However it is certain that by his actions and technologies, man exerts a strong influence on the environment and evolution. It is also certain that the tempos of cultural and evolutionary change have different orders of magnitude. To reduce this gap and tune in to our biology could be crucial for the survival of the species.

The approach of Wilson and the sociobiologists however tends to be reductionist. The replies of the sociobiologists to the criticism that sociobiology offers only a limited explanation of complex social situations, usually centre on the fact that the problem has too many variables but that it would be enough to further reduce the situation, refine the tools and do a little more research to solve the problem. In other words, says Wilson, if the forces which act between two bodies are known, those acting in a more complex situation (between many bodies) are the same forces. The same is true for cells, biological molecules and living beings. This point of view is decidedly reductionist. *The new alliance*[9] has some very interesting pages on this subject. Prigogine and Stengers claim that the discovery of dissipative structures brings us closer to the specificity of life and avoids the old conflict between reductionists and antireductionists. They quote a beautiful passage of Bergson, "An artist has painted a figure on a canvas. We can imitate his painting with multicoloured mosaic tesserae. We will reproduce his curves and subtleties better, the smaller, more numerous and variegated our tesserae. But we would need an infinite number of infinitely small elements with infinite colour gradation to make an exact equivalent of this picture. However the artist's conception was simple. He recreated it as a unit on the canvas and it is all the more perfect for being the projection of an indivisible intuition." We need to reach a more balanced "conception of the roles of the respective macroscopic parts and parameters which define the system as a whole". Microscopic and macroscopic must be reconciled. We need to study biological phenomena in terms of self-organisation capable of rendering globally coherent the behaviour of the individual parts.

Prigogine and Stengers continue that Michel Serres often writes of the respect which peasants and sailors have towards the world in which they live. They know that there is a natural time scale and that the development of living creatures, a process of autonomous transformation which the Greeks call *physis*, cannot be juggled. Once we discover nature in the sense of *physis*, we may also begin to understand the complexity of the problems faced by the social sciences. Once we learn the respect for nature that physical theory imposes, we must also learn to respect other intellectual approaches. We must learn not to judge the different forms of knowledge, practice and culture of human societies but to receive them and establish new channels of communication. Only in this way can we face the unprecedented requests of our time.

THE DEMOGRAPHIC PROBLEM
AND THE LOSS OF ESTRUS

In *The limits to growth*,[17] the requirements and availability of agricultural land are analysed up to the year 2100. The study shows that even if the society decided to undertake cultivating new land or increasing the productivity of land already cultivated, the population increase would still lead to a world food crisis within 40-70 years. The report goes on to say that it is not certain exactly how many people the Earth can nourish, because this depends on many alternatives. The limit of population growth is therefore a fundamental limit to be applied for the survival of the human species. World population is growing exponentially: 500 million people in 1650, 1.5 billion in 1900, three billion in 1950, six billion today. For millions of years the population was much less than the production capacity of the planet. For the first time in human history, saturation is approached as predicted by Malthus. Perhaps the Earth can still produce enough food for the six billion people, but the crisis will certainly be dramatic, in terms of food, energy and resources if the population doubles to 12 billion people as predicted in the next 40-50 years. This is a tiny interval on the biological timescale, and will involve our children and grandchildren.

When Umberto Colombo and Giuseppe Turani gave their book the name *The second planet*[26], they were referring to the four billion new inhabitants that the Earth will have to support by 2030. "This time," they write, "the population will double not because it is drawn by some force, but simply under the impetus of its numbers. We will have to make room for as many people as have appeared since the

---

[26]   U. Colombo and G. Turani, *Il secondo pianeta*, Mondadori, Milan, 1982.

origin of man. In order to do this we have barely 50 years, five decades, a couple of generations." They add "Never before has the world seemed so incapable of solving its problems and making room for new inhabitants. There is a shortage of food, energy, employment."

The population problem is a new historic variable of prime importance: when man became a farmer the world population was several hundred thousand (less than the population of a small city); at the time of Christ it was certainly less than 500 million. Historical and biological tempos speak very different languages. Even recent history is of no help in predicting the future on the basis of the emerging parameters. There is little time to learn the new language and culture of biological survival.

One of Lester Brown's books is entitled *Twenty-ninth day* and begins with a charming anecdote which goes something like this. There is a water lily leaf in a pond. Every day the number doubles: two leaves on the second day, four leaves on the third day, eight leaves and so on. If the pond is entirely covered in leaves on day thirty, when was it only half covered? Answer: On the twenty-ninth day. The anecdote is used by teachers in France to give children an idea of exponential growth. The water lily pond is our planet with its six billion people, and today it is perhaps more than half covered. In the next generation it could be completely covered. Here and there groups of leaves thicken at the edges of the pond, presaging the approaching event. Will these signs escape us or be wrongly interpreted, so that we cannot adapt our way of life and our models of reproduction in time? The thirtieth day does not offer the chance of survival. The only way to avoid it is to limit growth (population, energy etc) and acquire 'species consciousness'.

Obviously things are not so simple and the population problem has complicated relationships in politics, sociology, economics, biology, psychology and so forth. We shall try to analyse certain points with the intention of stimulating research and reflection. Let us start with the idea that a Swiss person consumes as much as 40 Somalians. In Europe we may consider that because we are approaching zero growth, population increase is not our problem: as if the crisis merely depended on the number of people and not on food, energy and resources. The population problem also concerns industrialised countries which plunder the food, energy and resources of the whole planet. Another birth in a western country is much more of a burden than 40 births in the third world. To speak of population growth does not make sense unless it is linked to the problem of production and distribution of resources on a world scale.

Paradoxically, the population increase is also linked to scientific progress in the field of medicine. The population increases not only because there are new births but also because the average lifespan is increasing: 30 years in 1650, over 50 years today. Obviously this does not mean that medical science should not be practised, but it is an extra factor that should be considered in predictions of population increase and in demographic and resources programming.

Population increase has been described as 'a human bomb that menaces the planet' and it is very difficult to predict the behaviour of man or of different populations in the face of absolute food and energy scarcity. In the next chapter we shall examine the relationships between this problem and war.

Bringing children into the world is often a response to precarious conditions of health and food supply: more children so that at least one will survive. The population problem cannot be divorced from its social and economic context. It is linked to the state of nutrition, health, education, elevation of the female condition and consciousness of the advantages that birth control can bring in terms of social and individual well-being.

China recently formulated a new approach to the relationship population-resources. Chinese demographic policy envisages birth control substantially based on social and moral custom. The Chinese approach has many contradictions and even certain apparently 'reactionary' solutions, but it is a far-reaching experiment worthy of serious attention.

The relationships between individual and social behaviour become complicated. The role of women assumes revolutionary significance for the first time in biological evolution. There is no doubt that reproduction is the essence of our fitness, of our biological success. But today our 'species consciousness' tells us that population increase can be the principal cause of the disappearance of the species, the cause of biological failure. The choice of opposing population increase means to go against evolution, or else it is an absolute novelty in the evolutionary field. We may have made modest discoveries of the relationships between environment and biology since Darwin, but nothing is known about the reciprocal adaptations of biology and culture. According to Barash, neither the male nor the female are complete in themselves and each is of equal evolutionary value. The biological behaviour of males and females is necessarily different. At this moment in history woman is becoming the protagonist of choices which have distant roots in a different biological history and in a different culture (or in opposition to an imposed culture) and in complicated feedback mechanisms between her culture and biology. Just as there is no doubt that the tendency to procreate and nurture her young is written into woman's DNA, it is also clear that woman's choices can favour the opposite tendency, slowing evolution and favouring the type of behaviour that furthers the survival of the species.

The evolutionary novelty may consist in the fact that in human females sexual activity and reproduction have not been automatically linked for some time. This variation of biological behaviour goes under the name of loss of estrus, and it is a true biological revolution. Giovanni Cesareo, in *The female con(tra)di(c)tion*[27] quotes Margaret Mead as saying that unlike female monkeys, the human female has acquired control of her sexuality. Cesareo adds that the loss of estrus marks a change in the sexual behaviour of the female and inaugurates the new sexuality of woman.

The loss of estrus or at least flexibility in sexual behaviour have also been noted in other primates (orangutans, chimpanzees, langurs, red monkeys). Many hypotheses have been put forward to explain the loss of estrus in the species Homo sapiens. They range from the most widely accepted (rather reductionist, in my opinion) which links the continuous sexual receptivity of the woman with the development of

---

[27]   G. Cesareo, *La con(tra)ddizione femminile*, Editori Riuniti, Rome 1977.

the bond of the couple and the greater necessity of male assistance in raising children, to the least likely (in my opinion) of Nancy Burley who proposes hidden ovulation because of the female desire to avoid pregnancy. Personally I am convinced that the problem needs to be studied more deeply and that only a sociobiological approach which takes cultural and biological behaviour into account can shed more light on the subject. I would like to think it is due to acquired social conscience in the female which tends to slow down the evolutionary process in order to ensure better quality of life for few children, or to a completely new form of biological behaviour which has acquired 'species consciousness' information (a larger number of children is to the detriment of the survival of the species) by biological and cultural interaction.

The problem of population increase shifts all the scenery, modifying economic, social, political and technological parameters. Unless population growth is slowed by man, it will irreversibly damage our quality of life. The population-environment relationship is a mathematical equation in two unknowns which has no solution. The industrialised countries continue to devour the resources of the third world countries, destroying the very basis of the biological environment. Coffee, tea and sugarcane occupy land required to feed millions of people. Africa witnesses increased desertification and erosion and the fastest population increase ever seen in any continent. Its per capita food production has dropped since 1970. Latin America allowed its forests to be cleared at a rate of 22 hectares per minute to create pastures to provide meat for the US.

Nor does another equation in only two unknowns have solutions: the population-resources relationship. Six percent of the world population consumes one third of world resources. It is therefore an illusion to propose the development model of industrialised countries for the third world. With the other two thirds of the resources, at most 18 per cent of the population could be brought to western levels, relegating the other 82 per cent to starvation.

Because the advanced industrialised countries consume ten times as much energy, food and resources as third and fourth world countries, every new birth in the west deprives ten people in poor countries of the possibility of life. Yet to defend the birth rate at all costs (even when a couple already has two children) and to obstruct birth control is known as 'the life movement'. Instead of anachronistic battles against the use of contraceptives, the Churches ought to add a commandment: do not have more than two children.

Because population control is a matter of survival or extinction for the species Homo sapiens, and hunger and war are inevitably implicated, precise choices must be made. The protagonists of these choices should be women in their new social roles and the new movements (ecological, pacifist etc.) which assign prime value to the quality of life. The alternative is to continue to think that the sun circles the earth instead of vice versa. Obscurantism never pays in the long run (Galileo docet): biological reality is staring us in the face.

ECOLOGY, THE SCIENCE OF COMPLEXITY

In 1848, 99 per cent of the birch moths (Biston betularia) in Manchester were white and one per cent black. The latter were easy prey, standing out clearly on the white lichen covered birch trunks. Then came the era of coal: pollution killed the lichens and by 1895 90 per cent of the birch moths in Manchester were black. The 10 per cent of white moths were in turn easy prey on the blackened birch trunks. Pollution imposed a direction on natural selection: the fittest survived and multiplied and the complicated interaction of mutations and environmental changes dictated the rules of Darwinian evolution. In cleaner places, for example Dorset or the Kensington Gardens and Hyde Park, the white birches only made life dangerous for the black moths.

Natural selection is inexorable: it means the survival of the fittest in changed environmental conditions, not adaptation to the environment with ad hoc mutations. But for man, this inexorability does not mean that we cast ourselves to the events of fate. For example coal pollution could be reduced, the particles eliminated, restoring serenity to the white moths. Obviously this is not the main problem, but a similar situation could involve the survival of our species, our health and that of our children. Man knows about evolution and within certain limits, can slow down the process and not allow the collapse of our complex environment, so essential for our life. Such action presupposes profound understanding of the biological mechanisms and the complex laws of ecology. The science of biology must be the basis of all our choices now that our economic, social and technological sciences have failed us and revealed themselves impotent against the now rapid, planetary scale changes. Astrophysics cannot be studied with the units and instruments of the civil engineer, just as the 2000s cannot be tackled with sciences which do not have large scales of time and space among their tools.

Much of what is described in Chapter 3 presupposes a knowledge of ecology, but it is obviously useless to repeat here a detailed description of the laws of ecology that can be found in the texts cited in the bibliography. In this Appendix we only want to deal with two points more relevant to the theme of this book: (a) the concepts of 'complexity' and 'stability' in ecology and (b) the fact that failure to conserve the rich genetic diversity of our planet means to sacrifice the survival of the species Homo sapiens.

Complexity/stability. The first point to emphasise is that a given word has different meanings in different disciplines. In thermodynamics the state of equilibrium is characterised by no change in the free energy function ($\Delta F = 0$) and the preferred state is that with minimum free energy. Minimum free energy in turn is determined by maximum entropy. Maximum entropy means maximum disorder and not maximum complexity; a 'complex' biological structure arises by processes which create order: minimum entropy within but entropy discharge without. The trend towards maximum entropy is a trend towards maximum disorder, towards the *'thermal death'* of the system: exactly the opposite to the biological *'complexity'* and *'stability'* which is life. The concept of equilibrium in thermodynamics thus has nothing to do with a stable ecosystem.

The words *'complexity'* and *'stability'* as used to describe a society or a method of production (for example the complexity of a technology or a city transport system, or the political stability of a country) are even more divergent and may even assume opposite meanings.

In biology, there seem to be at least two types of complexity: that of the single individual and that of the ecosystem. The second requires complicated definitions which take into account the number of species, their functions, the relations between species and their functions etc. One of the few certainties in this field is that the destruction of the diversity of our genetic heritage and the complexity of the ecosystem leads to processes of instability: damage to the environment, the uncontrolled increase of certain species and the destruction of whole ecosystems caused by the 'simplifying' action of man are plain to see. We can therefore say that a decrease in the complexity of an ecosystem generally leads to a decrease in its stability. In biology the tendency of an ecosystem to settle in a certain state with populations in fairly constant ratios, cooperative functions and minimum entropy production, can also be observed.

On the other hand, in evolution we observe a tendency towards increasingly complex individual species with increased entropy in the environment. I do not think that we can correlate complexity with stability in this case: a more complex individual is not normally more stable than a simple individual. Furthermore, a faster rate of evolution produces more entropy and causes destabilization of the life system in the long run. Again we encounter the concept of time (speed) and its relationship to complexity and stability: *to slow down the evolution of the individual and favour the stability of the ecosystem means not to further individual complexity but to favour the continuance of ecosystem complexity.* This might be the correct choice for the survival of the life system. The words 'slow down' and 'continuance' mean that time is included in the definitions of 'complexity' and 'stability'.

The last point is that the introduction of a species such as man, who has many relationships with many species, into a natural system, does not necessarily increase the complexity of the system. To the contrary it may mean simplification. A merely numerical definition of system complexity (for example based on number of interactions) is incorrect for two reasons. Man escapes many of the effects and consequences of the ecosystem and cannot therefore be measured with the same units. Secondly, in order to adapt to the environment, man has himself invented 'mutations', namely technological tools which enable him to falsify biological time, *'deceiving'* the other species and placing himself in a special relationship to the functions of the ecosystem. This is why the stability or destruction of the equilibrium of the planet Earth depend finally upon man's choices. If man decides to destroy himself, he will take with him all other living species.

So in order to work for stability and survival we must know biological reality and the rules of the game. These require an understanding of complexity and the recomposition of different disciplines. They require knowledge of ecology, the science of complexity.

A genetic library for agriculture. In a single night, during the warm humid summer of 1846, all the potato plants in Ireland were lost: two and a half million people died of hunger and another two million emigrated. The genetic basis of the Irish potato had become so restricted that it no longer had defences against a certain fungal disease. The genetic constitution of insects, fungi and viruses continually varies under the heavy pressure of chemical agents. A genetically new pathogenic species can spread to the whole population of a genetically uniform plant. In many parts of the third world such uniform populations are few, but in countries with industrialised agriculture, a narrow genetic basis is the inevitable result of the Faustian pact signed to obtain fast growing, high yield seeds. The US potato crop depends largely on only three varieties; the pea crop on only two. The same is true for wheat, soy and maize; all Brazilian coffee descends from a single plant.

Modern agricultural history is full of catastrophes of this kind: in 1860 phylloxera destroyed nearly all the vines in Europe and in 1958, 75 per cent of the durum and 25 per cent of the other wheat crops were destroyed.

Ten thousand years ago, according to the estimates of paleo-anthropologists, the world population was five million. These hunter-collectors had 25 square kilometers per head and exploited 5000 different types of plants for food. Today, with over 6000 million people, the population density is 30 persons per square kilometer and there are only 150 plants used for food in world commerce. In the meantime, botanists of the University of Texas estimate that 30-70 per cent of the Earth's plants will become extinct in the next 100 years and that a million species will disappear in the next 30 years. Every plant species which disappears takes with it 10 to 30 animal species, dependent upon it for food. These alarming figures give a measure of the erosion of genetic diversity. This process has accelerated in the last 30 years under the influence of the new agricultural policy (the so-called green revolution) exported by the US. In North America, 85 per cent of the genetic patrimony existing at the turn of the century has perished.

Certain American scientists are attempting to combat this tendency by creating a genetic library for agriculture. The germ plasm or genetic plasma of thousands of varieties of plants is stored with the aim of conserving the present genetic diversity of food crops.

During the ice ages only a few regions escaped the effects of the cold and succeeded in conserving their botanic patrimony with its relative genetic diversity. Today these areas are known as Vavilov centres after the Russian botanist who discovered them at the beginning of the century. They are all in third world countries. Thus we have the situation of a northern hemisphere with abundant food and no hunger problem, and a south rich in genetic diversity. The germ plasm is collected from these areas, but the species cannot be stored for thousands of years, and in storage they do not evolve as do their enemies in the real world. There are now even private gene banks owned by the multinational companies that produce herbicides and pesticides for agriculture. These varieties are subjected to forced genetic selection to produce new high yield hybrids. Developing countries are denied access to these banks and remain impotent spectators of the drain of germ plasm from their territories by the industrialised countries.

The protection of specific regions having a particular genetic heritage should therefore be a priority in international agricultural and environmental planning. Researchers and governments should commit themselves in this direction. The immense diversity of the genetic material is necessary for the survival of food plants and thus of man. The best insurance against destruction of food resources for the future is the continued existence of wild and primitive varieties of plants in the world. The diversity of the genetic heritage, the complexity of the ecosystem, must not be lost. *Biodiversity is the core of sustainable development.*

CHAPTER 4

MAYA, MAIZE, MALTHUS AND MARX:
A KEY TO THE PHENOMENON OF WAR

*G*uatemala: *a muddy track in the Petén jungle, a day and a half from the border between Honduras and Belize; the incredible sight of the pyramids of the Maya temples at Tikal.*

From the top of the Temple of the Giant Jaguar there is jungle as far as the eye can see. The photographs in the splendid little archeological museum of Tikal show that the temples were completely overgrown by plants when they were discovered. At Palenque in Mexican Yucatán and Copán in Honduras, there is the same vegetation: many marvellous little Mayan cities abandoned almost intact in the middle of the jungle. Archeological studies have revealed that each centre belongs to a different period: the Maya began to clear the forest and farm the land and gradually extended the cultivation of maize, the staple of Central America. In the sacred Popol Vuh of the Mayas, we read that the flesh of man was made from yellow and white maize, and his arms and legs of maize dough; the flesh of the fathers consisted only of maize dough.

Concentric fields of maize surrounded the village nucleus which was the place of ceremonies, administration and games. The farmed land lost its fertility or was insufficient and the Mayas were forced to farm poorer soils at greater distances from the centre. (The diminishing productivity of land is the first fundamental hypothesis of Malthus.) When the distances became too great, the centre was abandoned and returned to the jungle and a new town was built elsewhere. The farmer population was therefore, in a sense, also nomadic, like the hunter-gatherer peoples always in search of new land rich in game and fruits.

Although the first farming people made more efficient use of the land's resources than the hunters (which is why they overcame and superceded them), they too had to seek new territory. This system worked well for as long as there was practically unlimited land. The situation changed with population increase which was particularly marked in prosperous peoples. The agricultural population increased more than that of the nomads. Unlimited reproductive pressure is the second hypothesis of Malthus.

Once resources began to be scarce, wars for the conquest of land broke out. In this struggle which was the logical consequence of demographic pressure and the shortage of resources, Darwin saw the very basis for biological evolution. Conditions for evolutionary selection linked to the capacity to conquer new resources for survival, arose in this way.

At this point Marx rolls over in the grave and his accusation of 'reactionary tendencies' demands an answer. In *Entropia e potere*[1] Laura Conti writes:

"Marx rejected both the hypotheses of Malthus. Not only did he reject the idea of unlimited reproductive pressure, but he could not understand why Darwin, who had convinced him of biological evolution and whom he deeply admired, considered himself a disciple of Malthus. Malthus's idea that the human species had unlimited reproductive pressure seemed madness to Marx. In fact he thought that Darwin was unaware of the incongruence of ascribing this characteristic to a single species, as Malthus seemed to do. On the other hand, if other species also have this force, then that of man should match that of wheat, according to Marx, and thus man would never be without wheat. Curiously, it escaped him that man has the basic reproductive drive in common with the other species, but his capacity to realise it, and hence multiply it with the reproduction of

[1]   L. Conti in A. Baracca and E. Tiezzi, ed., *Entropia e potere*, pp. 25-34. CLUP/CLUED, Milan, 1981.

the survivors, is greater.... To the hypothesis of diminishing productivity of the earth, Marx retorted that soil fertility is a function of human work and is therefore destined to increase. But our everyday experience is that soil productivity tends to decrease...."

"Marx's argument is that the soil does not lose its original characteristics of fertility when cultivated, but incorporates in a lasting way the costs of the improvements made. Today we know that this is not true: most of the Fertile Crescent became a desert precisely because of the improvements made; the Amazon forests become a desert after very few crops."

Obviously Marx was lacking the concept of the productivity in terms of energy which decreases with increasing productivity in terms of labour, and all the thermodynamic literature on the phenomenon as a whole.

In *Questo pianeta*[2] Laura Conti writes:

"If we want reassurance about the survival of a descendant of ours different from us, I must admit that I am blindly conservative on this point: I would rather man remained as he is, with all his virtues and faults. I certainly would not refuse extra virtues, or perhaps even different faults, but I wouldn't like him to lose the ones he has. If they tell me that some of our descendants will have six hands and cat's eyes, I have no objection but I hope we don't lose the model that has been passed down for one or two million years. Greater variability certainly seems an acceptable idea to me, but I refuse to collaborate with a method of selection which causes further evolution of the species as a consequence of the introduction of greater variability. I quite like the evolution which has gone on so far, but enough evolution now. It's time to stop. ...."

---

[2]   L. Conti, *Questo pianeta*, Editori Riuniti, Rome, 1983.

"With these extravagant expressions I have only expressed an anti-evolutionary attitude which is in all of us, especially people of the left. I am trying to demonstrate the counter-evolutionary effect of everything which we consider positive i.e. humanitarianism, democracy, progress, communism, socialism and anti-racism. In these different fields, everything that opposes politico-social conservatism is biological conservationism. The scientists and doctors who produced insulin are the unconscious instrument of conservation of the genes that determine the heredity of diabetes. Architects who design houses and buildings which allow handicapped people to lead as normal a life as possible, are the unconscious instruments (if the handicap is the consequence of a hereditary disease) of the continuation of the genes which determine the disease. Teachers who devise didactic methods aimed to give everyone an equal chance of success, are unknowingly conserving the genes which determine lack of intelligence. This link between social progress and biological conservationism is not generally realised by those who, precisely because they believe in natural selection, work to oppose it. The connection between biological evolution as the interpretation of what happens in biology and the conservation of DNA as a guide to political and social direction, is also mostly unconscious."

In the phenomenon of war, the roles of natural equilibrium, resources and food production are thus linked to the role of man's biological and social behaviour. An even more complicated relationship between science and war emerges from the above.

Here we must beware of the facile tendency to reductionism: the purely 'ethological' point of view or at least the part of ethology which attempts to reduce human behaviour to animal tendencies. We have seen that man, because he is a cultural being, runs serious risks when his reproductive pressure exceeds that of wheat. In other words he is the only living creature able to destroy his own species and the natural equilibrium by choice. There is no doubt at all that man has the potential to choose a path which leads to a dead end for himself and the entire natural environment. We only need to think of the existing nuclear deterrent or the capacity to modify climate. On the other hand, because he is a cultural being, he can avoid the dead end and make a responsible choice for the future of his species and the delicate equilibrium of the planet Earth. This is the goal of 'species consciousness', the optimism of Sartre's last appeal: "Sub-men, do not despair!"

Environmental deterioration and waste of resources continue to penalise social development. The gap between rich and poor countries and the scarcity of resources lead in the direction of an unacceptable imbalance, with an increased probability of war. Man has sufficient history and culture not to obey blindly the laws of nature like an animal. He can slow down evolution rather than accelerate it in order to promote social equality. He can slow entropy production with appropriate choices; he can oppose the unlimited reproductive pressure with conscious choices of birth control; he can work for an international society of peace opposed to the war equilibria. Ecology, the science of complexity, can beat ethology. The 'social conscience of species' can win over 'natural behaviour'.

The other dogma to defeat is the 'neutrality of science' which has always been subordinate to the economic and political system in which it developed. My friend Dan Kohl, Professor of Biology at Washington University, renamed the Apollo project and the whole American moon programme 'the Nero project'. He rightly commented that, just as only the ingenuous would swallow the story about the Roman emperor letting Rome burn while he feasted and sang, only the ingenuous believe in the investment of millions of dollars to visit the lunar desert. No one would finance, to such an extent, an expedition to study the most unexplored desert in the world. The Apollo-Nero project had other aims: military control of space, the study of spatial communications for the interception of missiles, political publicity and so on. Science is rife with Nero projects.

71

In order to understand this intricate system of roles and the connections between science and war, a word is required about the most important resource for man's survival: energy. For millions of years man used renewable forms of energy: the sun, trees etc. Not long ago (in terms of biological history) man discovered non-renewable sources of energy (coal, oil, uranium) and this marked the beginning of a process that had two negative effects: (1) the discharge into the environment of heat and combustion products in an extremely short biological time (one to two generations) with respect to the time it took these resources to accumulate (millions of years) without giving the natural system the possibility to recover; (2) the input into the economy of value taken from nature to the benefit of one or two generations, defrauding future generations of a common non-reproducible good and creating false welfare for an extremely short period in the history of man.

The characteristics of the 'model of energy waste' are common to industrialised western and eastern countries: there are nuclear power stations and energivorous industries both in the USA and Russia. In both cases the lines of economic and industrial development are based on the destruction of the environment and the plunder of resources. The logical consequence is an imperialist type relationship with weaker countries, leading to war and conquest.

To this damage is added the impossible promise of exporting 'unlimited industrial growth' and 'welfare' to the third and fourth world. If the populations of the third and fourth world attained our energy consumption, the resources of the planet would be finished in 20 years and environmental catastrophe would be inevitable. The shortage of energy resources (and all primary materials in general) has recently culminated in a race, without rules, for the exploitation and control of every tiny piece of land, and in the flare-up of combat and coups in countries all over the world. Let us not delude ourselves that the victims of the third world will continue to leave this exploitation unopposed. If this waste and environmental destruction is not reversed, we will find ourselves involved in a violent scrabble for the last gram of energy. In this sense, to work for ecological balance is to work for peace.

To sum up: (1) the resources of the planet are limited; (2) war is a consequence of competition for resources; (3) to avoid war it is necessary to change over to a model based on renewable resources and the conservation of the environment.

Out of the hat comes the white rabbit of nuclear energy, the panacea of modern technology that can repair all the leaks in the energy plumbing system. American economist Herman Daly sees things a little differently. He draws a parallel between our present situation and that of a man in a sealed room with a water tap running. When the room begins to fill with water, a sane man will turn off the tap. An abnormal man will set to work with sponges and buckets and will call for increased production of sponges and buckets. Not only does it seem that we have chosen the sponge and bucket approach, but we have chosen the nuclear sponge and bucket.

Let us examine the nuclear alternatives.

(a) *Energy from nuclear fission* using enriched uranium: again this is a non-renewable resource.[3] If the growth of all countries continues at the present rate, world reserves of uranium will not take long to finish. The cost of uranium and traditional nuclear power stations is rapidly increasing. Power production is centralized and intensive. The consequences are still the same: low labour intensity (unemployment), inflation, increasing gap between countries having the technology and poor countries, serious risks for man and the environment, a radioactive waste problem for future generations.

(b) *Nuclear fusion* is the opposite of fission (from the Latin 'findere', to break) and consists in uniting (from the Latin 'fondere', to fuse) nuclei: the hydrogen bomb is an example of nuclear fusion.

---

[3] U. Colombo and G. Turani, *Il secondo pianeta*, Mondadori, Milan, 1982. The authors report that known reserves of uranium are equivalent to 30 billion tep which is one third the value of known petroleum reserves. With the technology of normal reactors, uranium is not a very interesting source and will run out much sooner than those of oil and gas.

In the hydrogen bomb, the temperature necessary to obtain nuclear fusion (100 million degrees) is provided by the explosion of an atomic bomb. In order to obtain energy from nuclear fusion, a container capable of holding material at this temperature is required. In *Il secondo pianeta* Umberto Colombo and Giuseppe Turani comment:[3]

> "The most that has been achieved is to hold a very small quantity of material for a fraction of a second at 15 million degrees, when what is required is to do it with a large quantity of material, not at 15 but at 100 million degrees, and keep the reaction going for a long time... This is why it is not thought possible to produce energy by nuclear fusion for at least another 40-50 years... Most of the conventional nuclear technology to date is irrelevant. It is a question of starting from scratch and perfecting one of the most complex technologies that man has ever conceived. It is obvious that this technology cannot be within the possibilities of all countries... There is also no guarantee that these power stations can be kept 'clean'."

These first two nuclear alternatives are certainly not our rabbit, but there is a third nuclear alternative which promises energy over a longer period.

(c) *Fast breeder reactors* use a mixture of plutonium and uranium. Plutonium is an element that does not exist in nature and is infinitely more toxic than cyanide (a millionth of a gram is a lethal dose). Its radioactivity has a mean half life of 24,000 years, and a kilogram of plutonium dispersed in the environment has the potential to provoke 18,000 million cases of lung cancer. To make up for this, its price on the black market is very high, much higher per gram than gold or heroin. The official journal of the American Chemical Society has reported mean annual thefts of plutonium of approximately 27 Kg.[4] Physics student John A. Phillips of the University of Princeton showed in his thesis that with a handful of plutonium and about $300 for assembly and electronics, a bomb one third the power of that of

Hiroshima can be built.[5] To quote Colombo and Turani again:

"Plutonium is extremely dangerous even if only used to *dirty* a conventional bomb... In the USA during the term of President Carter, the program for the construction of fast breeder reactors was blocked... All the fuel cycles of the different types of nuclear power stations are potentially 'proliferating'. In other words, they put those involved in a position to build a nuclear bomb."

This third nuclear alternative reveals the connection between the civil and military aspects of the nuclear question. The eminent English physicist, Amory Lovins, sustains that the necessary condition for nuclear non-proliferation is the elimination of nuclear energy production. He confirms that the cycles of nuclear fuel (uranium and plutonium) are so interchangeable and interdependent that it is impossible to speak of peaceful and military uses. A plutonium bomb, he adds, can be built in days or even hours, so that even the immediate discovery of theft is not sufficient guarantee that the appropriate measures can be taken.

Scientists, economists and journalists ask if it was really necessary for France to build her grandiose plutonium monster, the Superphénix, and for Italy and Germany to invest huge sums in this fast breeder reactor. The obvious reply is that Superphénix will produce enough high quality plutonium for 60 hydrogen bombs per year, or that the military sector needs fast reactors. Perhaps after 2000 years, Europe needed another 'Nero project'.

---

[4]   *Chemical and engineering news*, **55**, May 16, 1977.
[5]   See for example McPhee, J. *Il Nucleare tra guerra e pace*, Garzanti, Milan, 1983.

CHAPTER 5
ENERGY FROM THE SUN

I n physical terms, energy can be defined as the capacity of a body to perform work. Beyond this technical definition, we all have an intuitive idea of the meaning of energy. We only have to look around us. Everything is energy: from the food we eat to central heating; from the clothes we wear to transport; from the telephone to informatics; from plastics to chemical products. Energy is also information that we supply or obtain in a process. For example a tomato contains captured energy in many forms: solar energy, petroleum (fertilizers, insecticides, fuel and materials used for transformation), information (knowledge of farming methods), human work etc. Perhaps Jacques Prévert unknowingly makes a scientifically correct statement when he writes that when an old, old parrot brought him sunflower seeds, the sun came into his childhood prison. Thus there are many interchangeable forms of energy. The fact that one form can be tranformed into another enables natural resources to be used. Human work transforms them into manufactures.

Energy is therefore a fundamental part of a production process, and societies like Europe (in transition from an industrial to a post-industrial system) owe their development to the existence of energy sources. This is why energy today has assumed such importance for production-employment prospects and the problems of the environment. The transition of energy from non-renewable sources to clean renewable sources is probably the main problem to solve if humanity is to survive the energy crisis, environmental deterioration and economic problems which characterise the present age.

According to Commoner[1] these three aspects are closely related.

---

[1] B. Commoner, *The poverty of power*, Alfred A.Knopf, New York, 1976.

Environmental problems arise because energy is extracted without regard for its replacement. We only have to consider that two generations of man have practically depleted resources of coal and oil which took thousands of years to accumulate. To burn such a quantity of energy in such a brief historical period is to compromise ecosystem equilibria. Chapter 6 examines the environmental consequences of nuclear energy, and Chapter 7 the effect of carbon dioxide discharge into the atmosphere from the combustion of coal and oil. The complex relationships between the energy and economic crises can be better understood if we introduce the concept of 'productivity'.

Productivity in terms of labour is the quantity of value added per hour of work. Productivity in terms of capital is the quantity of value added per unit of fixed capital invested. The investments of recent decades have favoured the first type of productivity rather than the second, i.e. preference has been given to the large scale investment of capital in machines (with low capital productivity) to produce goods without the need for much manpower (high labour productivity). Hence the GNP grows but unemployment increases and there is a dramatic drop in the capital available to sustain economic growth or develop new forms of energy production, such as renewable resources. Nor are low-energy, labour-intensive sectors (which have little capacity to pollute) favoured. This leads to a heavy drop in the productivity of energy, i.e. the economic efficiency of energy transformation. The choice of nuclear power is an energy-intensive – capital-intensive choice with low energy and capital productivity (high waste) and reduced employment. The choice of capital-intensive production is convenient in that the use of machines decreases worker control over the production cycle. Technologies of increasing complexity deprive the worker of any say in the processes or risks of energy production. The cost of the energy produced will undoubtedly be less but the initial capital investment will be extremely high and there will be very little need for manpower. It can also be added that to continue to favour very high labour productivity and low capital productivity, is to promote the more noble forms of energy (i.e. those having a higher technological value) like electricity. These forms involve an absurdly high degree of waste.

An example of this waste is the electric domestic water heater. In the thermoelectric power station, water is heated by burning coal or oil to produce steam which drives a turbine connected to a generator. The electricity produced enters the national grid and is carried hundreds of kilometres (undergoing many losses) in order to again heat some water for someone's bath! This is what Commoner called a 'thermodynamic slaughter'.

The simultaneous occurrence of three crises ought to suggest energy choices which take the emphasis away from electricity. In modern technological society, two sectors account for a large slice of energy consumption: low-temperature uses (about one third) and transport (about one quarter to one third). Chapter 8 details some aspects of the energy question in the field of transport. The possibility of modern agriculture and advanced agrotechnologies for the production of renewable energy for transport are discussed. Here it suffices to say that the gross energy waste of road transport could be limited by reorganisation in favour of rail and water (rivers, canals, sea).

The low temperature sector is particularly interesting. It includes civil and industrial heating, heating of public buildings, preheated water for industrial use, washing etc. Investment in low temperature forms of energy means abandoning current ideas about productivity in favour of productivity in terms of energy. It also means investing in low grade, high yield energy and favouring the productivity of capital. There is no need for expensive machinery but rather for widely distributed small enterprises which can build the necessary equipment with low capital investment. Finally, this alternative will lower the productivity in terms of labour, shifting profit to productivity in terms of energy and capital. In this way it answers worker demands and creates new employment possibilities. The employment implications of energy choices ought to be a high priority point.

The present day energy picture is one of jugglers and tight-rope programmes which regularly prove to be biased or inflated and to ignore the capacity of the natural environment to annul the effects of our exaggerated energy consumption.

It is no longer true that economic growth and welfare go hand in hand with energy consumption. To the contrary, intelligent government of a modern society should be aware that resources are limited and impose on the system of production a policy of saving favouring low energy production, especially if based on renewable resources.

Inversion of the traditional growth of demand, namely the decrease in energy consumption which seemed a dreamers' utopia until only a few years ago, is becoming more and more feasible.

A new policy of energy saving and strategies of economic progress based on low growth of energy demand (taking energy as a means rather than an index of welfare) could be determinant in establishing a new and more correct order of the world economy and in closing the gap between the northern and southern hemispheres. It is well known that the per capita energy consumption in industrialised countries is about ten times the subsistence level whereas in the other countries it does not even reach this minimum level. It has been calculated that every inhabitant of North America (6 per cent of the world population) consumes an average of 62 barrels of crude oil per annum. In Eastern Europe (nine per cent of the world population) the figure was 28 and in Western Europe, Japan, Australia, New Zealand and South Africa (14 per cent of the world population) 22 barrels per year. These figures are food for thought, especially when compared with the average per capita energy consumption of the rest of the planet: Latin America, Africa, Asia (71 per cent of the world population) three barrels per year!

The underdevelopment and privation of the poor regions are likely to increase unless resources can be more fairly distributed. The industrialised countries have a fundamental role to play. Besides intensifying relationships with developing countries, they can reduce the serious tensions on the petroleum market. Unfortunately these steps have not yet been taken. The governments of the industrialised countries have never gone beyond solemn declarations.

There is an opportunity to counteract entropy increase and to increase energy availability by various forms of exchange of energy with the sun. From the thermodynamic point of view, the Earth is a closed system: it cannot exchange material, but it can exchange energy by capturing that of the sun.

As long as the sun continues to shine (long biological times), its energy can be captured in many forms: electricity (photovoltaic), heat, wind, bioenergy (biomass, biogas etc.), waves, tides, hydro etc. Each year the sun delivers 5.6×10^{24} joules of energy to the earth. This is 10,000 times the world's energy consumption (about 3×10^{20} joules per year).

Solar energy in its different forms could go a long way towards solving the problems of pollution typical of non-renewable resources (oil, coal, nuclear), including those due to the carbon dioxide produced by the combustion of fossil fuels.

BIOMASS AND GREEN PETROL

The energy equivalent of the agricultural and forestry biomass produced on the planet in a year is about 3×10^{21} joules, which is ten times the present yearly world energy consumption. It is a renewable resource. It is clean, undepletable and greatly under-exploited. There are two very important uses: fuel for cogeneration in small-medium urban centres, and 'green' petrol (alcohol) which can make a notable contribution to transport, a form of consumption typical of industrial society.

Ethanol and butandiol can be distilled from different biomasses by various well-known means. Animal fodder, fertilizers and chemicals are coproducts.

Possible prime materials are agricultural residues such as wheat and rice straw, corn stalks, prunings, or forestry residues and urban waste. They can be supplemented by ad hoc crops on marginal land (e.g. topinambur) and by high energy crops (sugar beet, sweet sorghum) when these do not compete with food production.

The proposal of certain agricultural multinational companies to build giant energy consuming plants with polluting production cycles, to use up the wheat surplus is clearly not what is needed, apart from the ethical question posed by world hunger. The same is true of the proposals of certain refinery companies to produce chemical additives derived from petroleum (e.g. MTBE and MAS) having dubious environmental repercussions. In any case their long-term effect on the complex biological sphere is impossible to predict: tetraethyl lead is the classic example.

It is estimated that available dry biomass could supply enough ethanol to cover 12 per cent of Italian national petrol consumption. The cost per litre of the ethanol produced would depend on the technology used. The optimal plant size is of the order of five million litres per year of ethanol.

GEOTHERMAL ENERGY

The internal temperature of the earth has been found to increases by 1°C for every 33 metres of depth. This value is known as the mean geothermal gradient. Geothermal anomalies exist in some areas, where the gradient may be as high as 4°C every ten metres. Rainwater filtering into the ground may be heated and return to the surface via fractures and faults as superheated steam or steam mixed with hot water. Areas in which this phenomenon occurs are known as geothermal fields.

Three types of geothermal field may be exploited industrially: dry steam, wet steam and hot water. Only five geothermal sources of dry steam, one of which is Larderello in Italy, are known in the world. They can be tapped directly, the steam being sent to a turbine to generate electricity. The wet steam deposits consist of water above boiling point trapped under pressure. When vented to the surface it appears as steam mixed with boiling water. The vapour can be used to generate electricity and the hot water can be used for desalination, distance heating (teleheating), hot houses for plants and so on. Even low grade thermal deposits (50-90°C) can be used for such applications.

While people all over the world are trying to discover

alternative energy sources, in Iceland geothermal hot water has been used industrially and for heating for many years. Heating in Reykyavik, for example, is all geothermal and application to green houses enables even bananas to be grown and to ripen.

In Japan the applications include the use of hot water for fish farming, laundries and domestic heating. Another possibility is the use of the geothermal energy of hot dry rocks. These constitute an immense reserve which if exploited correctly could meet the energy requirements of the entire globe.

WIND ENERGY

This type of energy exploits the rotation imparted by the wind to 'windmills' which have long been used to pump artesian water. Modern plants convert the mechanical energy of the shaft into electricity. The world trend is to increase unit power from the 100-200 kW plants already available and well proven, up to over 2 MW. Wind parks are the most economically feasible solution. The use of wind generators depends greatly on the amount of wind: good results are obtained with regular winds exceeding 4-5 m/sec, and these occur for example in about 15 per cent of Italian territory. Wind generators of about 1 MW have been tested with good results. The wind kW can almost compete with that produced from fuel oil.

The irregularity of winds is less of a problem when the electric network links meteorologically different areas. In Italy small wind generators are produced and with some good technological and industrial improvements, ten per cent of Italian needs could be met without distribution problems.

PHOTOVOLTAIC ENERGY

Photovoltaic technology utilises a practically unlimited energy flux. Solar radiation in countries like Italy reaches 1 kW per square metre and today, a 0.3 × 1 m panel can provide 40 watts of power. It is there-fore correct to regard photovoltaic energy as an important component of a different energy future. A feature of this form is the direct conversion of solar radiation into electricity, without mechanical intermediaries or pollution. For it to be an economic possibility, much technological innovation is required, and a reduction in costs due to mass production. Although photovoltaic energy did not achieve the success predicted for it in the eighties, it rightly continues to sustain much hope. Its competitiveness depends on the particular application. It is already competitive in inaccessible areas and isolated unmanned stations.

LOW TEMPERATURE SOLAR ENERGY

This form of energy is suitable for heating water and buildings. At the moment its use is limited to substitution of electric water heaters, but its adoption is impeded by problems of certification of quality, high installation costs and competition from methane.

COAL AND FUEL CELLS

There are still enormous deposits of coal and it is still the cheapest form of energy, but at every step, its use causes serious environmental problems. When the environmental question arose, a series of environmental protection measures and technologies came into being. The result was a steep rise in cost of plant. The range of coal technologies is nearly infinite, and with careful choice, permits the use of coal in forms and sizes appropriate to the site. Hence transitional use may be made of this resource, even though it is non-renewable and produces carbon dioxide. For example there are 'fuel batteries' suitable for urban use, which have a high yield of electricity (50-60 per cent) and also produce heat. Present cells have a power of approximately a MW and work with hydrogen or a similar gas, produced from methane. It is soon hoped to produce it from coal. These cells are almost economically competitive.

HYDROELECTRIC POWER

Hydroelectric power stations produce clean, renewable, low running-cost energy but the approach is different from that to wind and photovoltaic power. The latter are largely new and open to innovation, whereas hydroelectricity is not a new resource. Here a known resource requires development with new criteria and consideration of the environment. In Italy, apart from reactivating power stations which have been closed down, there are two avenues, thanks to the possibility of automatic functioning and the production of low cost modular plants in small series. The first is a programme of private initiative favouring access to a local resource and promoting autonomy in energy terms, for example farms distant from the national electricity network, or small communities. The second, which will give greater overall results, is a programme of territorial planning which places hydroelectricity in the framework of hydrogeological management and renewed economic impetus to highland activities. This resource, which could supply five per cent of the national requirement (in Italy), needs to be included in a programme of reintegration of emarginated areas.

NATURAL GAS

Methane is the cleanest of the fossil fuels, its supply on the international market is growing and its price is about the same as that of fuel oil. Natural gas is the main fossil resource in Italy. Known deposits account for about 250 Mtep which could supply 35-40 per cent of Italian needs. In the framework of a policy of energy reform, it is therefore the best fossil resource for the next few decades.

From this point of view, natural gas is well suited to urban cogeneration-teleheating with suburb-size power stations. Some of the possible technologies, like the gas turbine-vapour turbine combination and fuel cell, are very interesting.

Italy has contracts with many methane producing countries (Holland, Russia, Arab countries). Consumption is expected to

TABLE 1: ESTIMATED COST OF DECOMMMISSIONING CLOSED NUCLEAR POWER STATIONS				
Owner/site	**Capacity** (megawatts)	**Estimated cost of decommissioning** (millions $ – 1985)	**Cost** ($/watt)	**Years of functioning**
US Atomic Energy Commission, Elk R.	22	14	0.58	1962-1968
UK Atomic Energy Authority, Sellafield	33	64	1.94	1963-1981
Pacific Gas & Electric, Humbolt Bay, Unit 3	65	55	0.85	1963-1976
US Dept. Energy, Shippingport	72	98	1.36	1957-1982
Commonwealth Edison Co., Dresden*	210	95	0.45	1960-1978
* decommissioning complete				

Sources: *Nuclear Public Utilities Commission*, San Francisco, 1985; *OECD Nuclear Energy Agency*, Paris, 1985; *Nucleonics Week*, 25th April 1985; *Public Citizen/Environmental Action*, Washington DC, 1985.

increase from 14.5 Mtep in 1973 and 27 Mtep in 1984 to 40 Mtep in 2020 and then decrease to 30 Mtep in 2050 following the trend of diminishing use of fossil fuels.

THE LIMITS OF THE NUCLEAR ALTERNATIVE

As everyone knows, the problems of nuclear power stations not only concern the reactor itself, but the whole cycle, from the mining of uranium to the reprocessing of irradiated fuel and the disposal of highly radioactive wastes. It is difficult to devise truly stable waste storage systems because the radioactive life of this material is of the order of tens of thousands of years. In countries like Sweden, failure to solve the problem of nuclear waste disposal played a large part in the decision to phase out these power stations.

Serious problems are also involved in the decommissioning of electronuclear power stations at the end of their life cycle, which is about twenty years. Table 1 gives estimates of the cost of decommissioning for several nuclear power stations in millions of dollars (1985). The second column gives cost per megawatt of the installation.

The cost of nuclear installations and of the nuclear kWh is another controversial question about which there are many misunderstandings. Table 2 shows the original estimates and the 1985 costs of

TABLE 2: INCREASE IN COST OF NUCLEAR POWER STATIONS		
	Estimated cost (billion dollars)	**Actual cost (1985)** (billion dollars)
Diablo Canyon (Calif.)	0.44	
Shoreham (NY)	0.25	4.4
Midland (Michigan)	0.25	4.4
Marble Hill (Indiana)	1.4	7

TABLE 3: COST OF ELECTRICITY PRODUCED BY NEW POWER STATIONS (cents per kWh, 1982 dollar)		
	1983	**1990 estimate**
Nuclear	10-12	14-16
Coal	5-7	8-10
Minihydroelectric	8-10	10-12
Cogeneration	4-6	4-6
Biomass	8-15	7-10
Wind energy	15-20	6-10
Photovoltaic conversion	50-100	10-20
Efficiency	1-2	3-5

Note that the cost of coal, mini-hydroelectric, cogeneration and biomass production are competitive with nuclear power.

construction of several US power stations. Table 3 gives 1983 figures and predictions for 1990 of the Worldwatch Institute (USA) for the cost of electricity produced in different ways. This table is reported for the purpose of historical comparison.

COGENERATION AND TELEHEATING.

Fifty percent of the energy of a thermoelectric power station (coal, oil) is lost as hot water. This energy in the form of heat could be used in centralized heating (teleheating) of towns and industries. This is known as cogeneration. It reduces the yield of electricity from 40 per cent to 35 per cent but permits 55 per cent of the energy lost as heat to be recovered. This brings the final efficiency to 90 per cent rather than 40 per cent. The first teleheating system was built in USA in 1877. Its use expanded until about the twenties when it was slowed by coal and oil producers. At the present time, the electrical authorities or municipal energy companies of hundreds of northern and eastern European towns supply not only electricity but also teleheating. The houses of six million inhabitants of Germany, nearly half the population of Sweden and five million people in Moscow are heated in this way.

CHAPTER 6
NUCLEAR ENERGY: FROM MEDIEVAL TECHNOCRACY TO FUTURE IMPERFECT

The final phase of the nuclear fuel cycle, namely the disposal of nuclear wastes, involves many risks. This material is highly radioactive and has a half-life of tens of thousands of years.

The first radioactive wastes were dumped in the deepest parts of the ocean without taking into account that the pressure at such depths would break the containers and release the contents. France and United Kingdom irresponsibly failed to take note of where the containers were sunk. Geologically stable sites (granites and basalts) devoid of water have since been preferred, however the half-life of the wastes are of the same order as the geological time scale, making it impossible to guarantee the stability of these sites. In the meantime some science fiction solutions have been thought of, for example wastes in orbit in satellites, launching wastes into space or onto the Moon.

Today no one has yet built even a model wastes containment system. The nuclear industry has never been much concerned about our health. It has only ever spent money to safeguard public health when forced to do so or when it feared economic consequences. At Gorleben in West Germany, research was undertaken as a result of a law making building permission for a reactor conditional to the demonstration that the wastes could be contained until their decay to low radioactivity levels. There is a similar law in Sweden and in two states of the US: California and Virginia. Since the law was passed, not one new reactor has been authorized in Sweden. The law was made as a result of protests by ecological and antinuclear groups.

It is said that all technologies involve risk, but the unsolved problem of radioactive wastes introduces a new variable: for the first time in the history of man a mortgage is placed on future generations. This risk, together with the obligation to decommission reactors after a few decades, is a social cost of unforeseeable size.

In the year 3000, in the course of marine research or excavation of some granite cave, an archaeologist might find a wonderful chest, an 'Etruscan tomb', full of radioactivity.

THE BIOLOGICAL CYCLE OF RADIOACTIVITY

The effects of radioactivity on living organisms are varied, and can be extremely detrimental to health. In order to understand them, specialised biological knowledge is needed. The manner in which radioactive elements discharged into the atmosphere, rivers and seas are distributed, is complex. Complex too, is their metabolism in different animal and plant species. Organs faced with the various radioactive chemical elements react in different ways. The path of these elements in the biosphere and food chain is also complex. They are taken up by our bodies and their effects may accumulate in time, with unpredictable results.

Every radionuclide has its own particular, largely unknown path from its source to man. For example the accumulation of different radioactive elements may range from one in water to 40,000 in fish, which is an important food item for many populations. This makes it difficult to consider schemes of 'safety threshold' so dear to supporters of nuclear power. Figure 3 shows how the various elements released by a nuclear power station may reach man. Radioactive discharge to the atmosphere constitutes a secondary risk to the man who breathes it because the concentration is low. The fundamental concept concerns the risk to those who are forced to live in contact with traces of radioactivity for many years. These may reach man via drinks and food, i.e. via the food chain. Fish feed on algae and plankton which accumulate radioactive elements from contaminated water; plants absorb wastes directly from the soil and air and are fed to domestic animals. These small quantities add up in time and harm man.

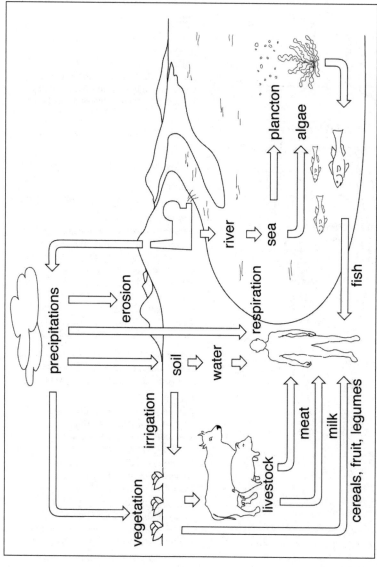

Figure 3: The biological cycle of radioactive elements from the nuclear power station to man.

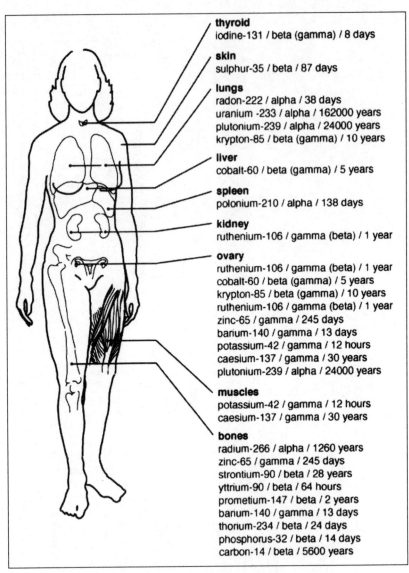

thyroid
iodine-131 / beta (gamma) / 8 days

skin
sulphur-35 / beta / 87 days

lungs
radon-222 / alpha / 38 days
uranium -233 / alpha / 162000 years
plutonium-239 / alpha / 24000 years
krypton-85 / beta (gamma) / 10 years

liver
cobalt-60 / beta (gamma) / 5 years

spleen
polonium-210 / alpha / 138 days

kidney
ruthenium-106 / gamma (beta) / 1 year

ovary
ruthenium-106 / gamma (beta) / 1 year
cobalt-60 / beta (gamma) / 5 years
krypton-85 / beta (gamma) / 10 years
ruthenium-106 / gamma (beta) / 1 year
zinc-65 / gamma / 245 days
barium-140 / gamma / 13 days
potassium-42 / gamma / 12 hours
caesium-137 / gamma / 30 years
plutonium-239 / alpha / 24000 years

muscles
potassium-42 / gamma / 12 hours
caesium-137 / gamma / 30 years

bones
radium-266 / alpha / 1260 years
zinc-65 / gamma / 245 days
strontium-90 / beta / 28 years
yttrium-90 / beta / 64 hours
prometium-147 / beta / 2 years
barium-140 / gamma / 13 days
thorium-234 / beta / 24 days
phosphorus-32 / beta / 14 days
carbon-14 / beta / 5600 years

Figure 4: The site of accumulation of various radioactive
elements entering the human body in the diet.
Each element has its own half-life and preferred organ.

Figure 4 shows the site of accumulation of various radioactive elements entering the human body in the diet. Each element has its own half-life and preferred organ. Iodine-131 concentrates in the thyroid which may be fatal in newborn infants; sulphur accumulates in the skin and may lead to skin cancer; cobalt builds up in the liver, and so on. The most delicate organs are the bones, where strontium replaces calcium and irradiates the marrow, and the ovaries, which may be subject to genetic mutations. The sex organs are attacked by all isotopes which emit gamma radiation.

The fact that the accumulation of radioactive elements varies according to organ and organism makes it meaningless to speak of radioactivity threshold.

Plutonium-239 concentrates in the gonads causing birth defects and malformations from the first generation. Every element (iodine-131, cobalt, krypton, ruthenium, zinc, barium, potassium) has its own destination in the body and takes part in different biochemical reactions, acting at different levels in different organs. There is a whole series of possible radioactive elements with extremely long half-lives (up to thousands of years) which can concentrate in the various organs, causing damage which may not become evident until much later. Others are more labile (e.g. iodine-131 has a half life of only eight days) and have immediate effects.

LOW DOSES, NATURAL BACKGROUND AND REPAIRING ENZYME SYSTEMS

During its normal function, every nuclear power station emits radiation and radioactive atoms. A single radioactive atom or radiation can damage the structure of a cell and its 'message-centre', the genetic code which regulates normal growth. If this part is damaged a single cell can multiply out of control causing the death of the organism. Radiations can modify the genetic message in the sperm or ovum, causing birth defects. People living near nuclear reactors (within a radius of 50 or even 100 Km) are exposed to low doses of radiation and radioactive atoms. Low doses over a period of time are more dangerous than the total in a single dose. Low doses over a long period can lead to tumours or mutations. A crucial point is that it is impossible to define a risk threshold for mutations and cancerogenous processes. This is due to the complexity of cell replication processes and the existence of random factors having cumulative and synergic effects, but the main reason is that the response varies from one individual to another. The most reasonable conclusion is therefore that any dose of radiation involves a risk of cancerogenesis or mutation for some part of the population. "Any dose of radiation is an overdose," says George Wald, Nobel Prizewinner in Medicine.

Nuclear supporters reply that the natural background of radio-activity is greater than the doses emitted by a reactor and that the risk involved in living in an area with a high natural background, or in moving from an area with a low to one with a higher background, is greater than the risk involved in living near a reactor. This argument is schematic, simplistic and nonsensical for the following four reasons.

(1) The effects of the natural background of radioactivity are not known. The first order of uncertainty concerns the type of investigation. A well-known article by M. Sakka[1] has shown that the studies performed so far are too broad and that only more pointed analysis can tell us anything about the carcinogenicity of natural radioactivity. The second order of uncertainty regards the genetic adaptation of populations (their natural selection) in relation to the metabolism of radioactive elements and the relationship between genetic capacity to repair damage (see point 3) and the effects of low doses of natural radiation. Only part of the population is at risk from exposure to low doses, i.e. people with no capacity for enzymatic repair. This part of the population may already have been selected out in areas with a high natural background level. For example some areas of Minas Gerais and Goiaz in Brazil have a very high background but the Indios population which has inhabited the area for thousands of years has evidently undergone natural genetic selection at the hands of the

---

[1]   M. Sakka, *Nippon acta radiologica*, **39**, p. 536, 1979.

environment, those with an inefficient enzyme system having been eliminated long ago.

Genetic selection could thus explain the low mortality in these parts of Brazil. No data is available on mass emigration. If 10 million Italians migrated to Brazil and 10 million Indios came to Italy, there might be a catastrophe of deaths from cancer among the Italians and from other pollution among the Indios.

However there is evidence, especially in recently colonized areas, of a relationship between chromosome changes, death from cancer and natural radioactivity levels. This has also been confirmed in uranium mineworkers. A recent study of groups of villages in Brittany has shown that the death rate from cancer in areas of granite with uranium is twice that in similar areas without uranium.

(2) It is impossible to estimate the risk of cancer from low doses of ionizing radiation. This statement was made by Prof. Charles E. Land of the Department of Epidemiology of the National Cancer Institute in Bethesda, USA, an authority on medical statistics. In an article in *Science,* Land[2] examines the reasons for the conflicting conclusion of studies of cancer risk from low doses of ionizing radiation. He first points out that we have very little information on the possible effects of a single rad[3] on man. In order to have a statistically significant sample, 10 million people would have to be studied. Otherwise data on the risks of exposure to high doses could be extrapolated to low doses. However this implies assumptions about the shape of the dose-response curve that give very different conclusions even when only a few of the parameters of the theoretical curve are unknown. In other words the choice of theoretical model affects the whole analysis. Land concludes that we have not the scientific tools necessary for an epidemiological study on populations exposed to low levels of radiation, hence any purported statistics on this relationship are likely to be incorrect.

---

[2]   C.E. Land, 'Estimating cancer risk from low doses of ionizing radiation', *Science,* **209**, 1980, 1197-1203.

[3]   A rad (rd) is a unit of measurement of the dose absorbed by an irradiated object. It is equal to 0.01 joule/kg of the irradiated object.

Another important point made by Land concerns the different behaviour of different forms of cancer. The same mathematical model cannot be used for the various types of radiation-induced cancer. In fact in the literature there are examples of conflicting epidemiological estimates due to the use of different models.

Land concludes that the choice of model can have more influence on the estimate of the risk of exposure to low doses than the data itself. He continues that it is unlikely that we can solve the problem of estimating cancer risk from exposure to low doses of radiation. Paradoxically, it seems that more information on low dose risk can be drawn from the study of populations exposed to medium to high doses than from populations only exposed to low doses, even when the samples are very large. This again emphasises the likelihood of drawing conclusions which are completely wrong.

(3) The genetic problem is linked to the fact that defence against low doses of radiation depends on the functioning of the enzyme repair system. If the enzymes of repair undergo a change in activity it leads to gaps in, or depression of the capacity to repair. Cells in such a condition are extremely sensitive to radiation, such as in subjects with genetic autoimmune diseases. The cells of patients with xeroderma pigmentosum, ataxia teleangiectasia, rheumatoid arthritis, lupus etc. are extremely sensitive to small doses of radiation; these cells can be regarded as human mutants and the patients have syndromes of chromosome breakage and predisposition to cancer. Different people have normal, partial or no repair capacity. The latter are exposed to risk from any level of radiation, unlike people with normal repair capacity.

These findings assume new significance in the light of recent discoveries on the mechanisms of cancer: a single unit of the neoplastic gene (mutant gene capable of leading the cells towards a condition of cancerous degeneration) differs from the healthy gene, and it is precisely this subunit that has the ability to cause the cells to degenerate. The four subunits of DNA are indicated by the letters C, G, A and T from the initials of their chemical names (cytosine, guanine, adenine and thymidine). Researchers discovered that a triplet of sub-units of the healthy gene, the triplet GGC (which gives the instruction for the formation of the aminoacid glycine) was replaced by the triplet GTC in the cancerous gene (which coded for the aminoacid valine). When the triplet GGC (healthy gene) occurred in the particular part of the gene investigated, transformations did not occur in cultivated cells in contact with the gene. These transformations occurred regularly in the presence of the triplet GTC. Hence only a tiny variation can transform a normal cell gene into one capable of causing cancer.

In a study on the repair of DNA and radioprotection[4], M. Quintiliani and O. Sapora added other important elements to the question: (a) "laboratory experience has shown that fractioning the dose and increasing the interval between the fractions, leads to a substantial reduction in efficiency (compared to when the same dose is given all at once), of most biological responses including cell death. This is because the repair systems repair the damage between one fraction and the next. For low doses and when the biological test is transformation rather than death, we have the opposite effect. This can be explained as a build-up of faulty repairs, because the most dramatic effect of radiation, i.e. cell death, does not occur at low radiation levels. Assuming that there is a relationship between trans-formation and mutation and that these processes are in turn related to the onset of carcinogenesis at an individual level, then such an observation in cells in culture could have practical repercussions. In this case, a linear extrapolation based on single doses would not be correct for estimating radiobiological risk. Lower exposure speeds are ten times more effective than higher speeds in triggering carcinogenic transformations in cells, because of the faulty tissue repair;" (b) "the analysis of DNA repair systems leads to the conclusion that chemical and physical agents are not only able to produce lethal dam-age in cells but can also cause potentially mutagenic or cancerogenic DNA damage. In certain human diseases, there is a high incidence of

---

[4]   M. Quintiliani and O. Sapora, 'Riparazione del DNA e radioprotezione', *Giornale Italiano di medicina del lavoro*, **3**, p. 175, 1981.

tumours. Three of these diseases, xeroderma pigmentosa, ataxia teleangiectasia and Fanconi's anaemia are associated with genetic defects of cell DNA repair systems"; (c) "these are rare diseases having an incidence of about one in 100,000 births. However there is still the problem of the heterozygotes which are much more frequent (more than 1/1000)."

It has been observed that even among apparently normal subjects, there are individuals whose cells are extremely sensitive to ionizing radiation. It is suspected that these persons might be heterozygotes for the main hereditary DNA repair diseases. People with these defects may long have been selected out in areas with a high natural background of radioactivity, but are certainly still present, for example, among the Italian population.

(4) Nuclear reactors discharge a series of radioactive elements into the environment, and thus into the food chain and the human organism. Some of them do not occur in nature or only in tiny concentrations or in different chemical states. The distribution of these elements in the various living organisms is extremely complex and unpredictable. The concentration of one of these 'unnatural' radioactive elements in a particular biological site at a given moment, may be thousands of times higher than in another nearby site (e.g. inside or outside a cell). This concentration depends on many chemical, physical and biological parameters which can change rapidly, even in a fraction of a second. As far as can be determined a posteriori or a mean concentration established for any element in a given system (e.g. a plant or animal), the result will always refer to an overall mean static situation and no information will be gained about the real situation 'in vivo' at a given time. Such information would require guided investigations using special techniques (e.g. nuclear spin relaxation) which would be lengthy and prohibitively expensive even for a single chemical element in a relatively simple natural system. It would thus be impossible in many parts of the human body. An example of the complexity of the problem: some elements can exchange rapidly from the inside to the outside of a cell, others

cannot; others only if helped or transported in, e.g. by a molecule or pharmaceutical, which may or may not be there at a given time; the same elements can also exchange energy in different forms with the biological environment or nearby molecules at a speed ranging from fractions of a second to days. Free radicals can further complicate the situation favouring or modifying the reactivity and energy exchanges of different chemical elements. The presence of a radioactive atom can create new types of free radicals which can cause cancer. It may therefore be said that: (a) the complexity of the situation described above is far from comparable with that of the natural background radioactivity and the presence in nature of certain radioactive elements; (b) man has a degree of 'genetic immunization' against the natural background and natural radioactive substances.

The complexity of immunological mechanisms (a field still little understood) and of the mechanisms of concentration of single cells or molecular sites introduces so many unknowns that it is clearly senseless to establish a safety threshold or to speak of 'dose'. As we have seen, epidemiological studies are not much help (point 2) and the latest discoveries on repair mechanisms (point 3) cast further shadows on what was hitherto believed to be true.[5] George Wald is right: any dose of radiation is an overdose.

---

[5] E.P. Radford, former president of the Committee on Biological Effects of Ionizing Radiation which published the BEIR reports, wrote in the December (1981) issue of *Technology review* that he was convinced that much of the technical information in BEIR III on cancer risk was obsolete and that new estimates were required. He went on to say that new data showed that cancer risk is substantially higher than indicated by BEIR III.

CHAPTER 7
TWO SEASONS INSTEAD OF FOUR

The atmosphere surrounding our Earth is essential for life. The chemical composition of the atmosphere changes with altitude and gases which occur in different proportions and sometimes in traces in the different strata, have very important roles in the equilibrium between the Earth and the Sun. They therefore affect the climate and the radiation reaching the Earth's surface.

The stratospheric ozone layer filters part of the ultraviolet component of the Sun's radiations which would otherwise damage our skin. The presence of a certain concentration of carbon dioxide maintains climate. A big increase in carbon dioxide would cause an increase in temperature and the desertification of the whole planet.

Like biological equilibria, the chemical equilibria of the different layers of the atmosphere are extremely sensitive to changes. Those induced by man can trigger irreversible processes and chain reactions. We will analyse examples such as tropical deforestation and atmospheric discharge of carbon dioxide from the massive combustion of fossil fuels in a time which is very brief on the natural time scale (millions of years were necessary for these fuels to form, and a large proportion of them has been burned in the space of two generations).

The US National Research Council and National Academy of Sciences, two bodies to which the foremost world experts on atmospheric sciences belong, published the following findings: (1) the qualities of the atmosphere which protect life are not limitless and can be seriously damaged by a rapidly growing world population; (2) because of the increasing complexity of our societies, we have reached a critical stage in our relationship with the atmospheric environment of the Earth; (3) food production depends on the climate and the negative impact of human activities on atmospheric quality is increasing; (4) two effects which deserve the most considerable attention are the destruction of stratospheric ozone by organic fluorochloro derivatives and atmospheric heating caused by the carbon dioxide generated by the combustion of oil, coal etc.

The Committee for Atmospheric Sciences of the National Research Council examined the state of the troposphere and stratosphere and confirmed that the continuation of these studies in order to predict the climatic effects of man's activities is a 'high national priority' in the next decades. The Committee stated that much data and knowledge was required. In the field of human impact on climate we have learned enough to sound the alarm but not enough to quantify the danger adequately.

Socioeconomically, the main problem caused by the increase of carbon dioxide is the effect of desertification and the partial melting of the polar ice caps on climate and agriculture. If the present trend in consumption of non-renewable fossil fuels continues, the quantity of carbon dioxide in the Earth's atmosphere will double in about fifty years. Scientists predict that such an increase could produce a greenhouse effect with the consequent significant increase in worldwide temperature. Changes of this kind may be of the same order of magnitude as those which separated the various geological epochs, upsetting the already precarious ecological equilibrium of the planet and damaging world food production.

The accumulation of carbon dioxide in the atmosphere poses an unusual political problem. Our culture and society developed in a period of almost absolute climatic stability. This is no longer so and the problem of carbon dioxide is rooted in the global use of energy resources and agriculture. The stakes are very high, there are many uncertainties and, politically, the only way to approach the problem is the difficult path of international cooperation.

THE CARBON CYCLE

The study of the carbon cycle in the biosphere is substantially the study of the interactions between living organisms and their environment. Along with the atmosphere, the actors in this cycle include the oceans, terrestrial and aquatic animals, dead organic matter, the biomass (especially plants) and fossil fuels.

Although carbon dioxide is only a minor part of the air, it is still essential for life on Earth for two reasons: (a) it constitutes food for plants which convert it to sugars and starch by photosynthesis; (b) it regulates the temperature of the surface of the Earth and hence the climate, by the greenhouse effect.

Photosynthesis was a milestone for life on Earth. Early organisms developed the capacity to use, aided by sunlight, carbon dioxide and water in their surroundings to build the organic molecules which they required for growth.

Seventy percent of the photosynthesis carried on is performed by land and the rest by marine plants. The biggest contribution comes from the forests of the tropical belt. Cultivated land only contributes three per cent. Carbon is fixed as carbon dioxide by plants and sooner or later is returned to the air as carbon dioxide or to the sea as organic material. This return occurs by two pathways: the respiration of consuming organisms (including man) and the action of the organisms which decompose dead organic matter and eventually return it to the mineral state. Photosynthesis and respiration cause daily fluctuations of carbon dioxide in the atmospheric reservoir.

In the sea, the carbon cycle is somewhat different and the photosynthesizing agent is phytoplankton. Carbon dioxide fixed in the surface layers of water initiates a downward flow of carbon. Organic sediment is used by the decomposing organisms of the ocean floor, again producing carbon dioxide which is partly absorbed in the depths of the sea and partly released to the atmosphere. The marine reservoir absorbs carbon dioxide from the other two systems (earth and atmosphere) by the rain cycle and surface absorption.

The potential of the ocean depths to absorb carbon dioxide is practically unlimited, except that the carbon is transferred from the surface to the deep waters extremely slowly. The speed of mixing with the deeper layers is of the order of thousands of years, whereas equilibrium between atmosphere and surface waters is only eight years.

Only the upper hundred metres of the ocean participate in this mixing of gases and exchange with the air. The ocean surface cannot absorb all the carbon dioxide produced by human activities. Exchange with the depths of the ocean is the process which limits the absorption of carbon dioxide, and, being very slow, the overall result is the accumulation of carbon dioxide in the atmosphere. In other words, the ocean depths cannot help to mitigate carbon dioxide build-up.

Human activities contribute to this build-up in two main ways: the combustion of fossil fuels (coal, oil, natural gas) and deforestation, especially of tropical rain forests.

Figure 5 is a simplified scheme of the carbon cycle, with particular reference to the exchange of carbon dioxide with the atmosphere. The terrestrial energy balance is based on this scheme: the intensity of incident sunlight, the percentage of energy absorbed and the quantity of heat radiated back into space determine the temperature of the Earth's surface. We shall see the importance of carbon dioxide and the greenhouse effect in this thermal equilibrium.

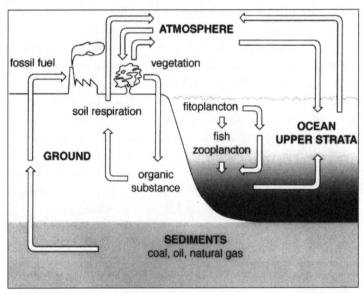

Figure 5: A simplified scheme of the carbon cycle.

Every year man discharges increasing quantities of carbon dioxide into the atmosphere. Except for three periods (the two world wars and the depression) the increase, from the industrial revolution to the present time, has been steeply continuous. Every ton of coal burned produces three tons of carbon dioxide. In 20 years there has been an increase in atmospheric carbon dioxide of 20 ppm (parts per million) which corresponds to 42 billion tons of coal. From the industrial revolution to the present time it has been calculated that approximately 100 billion tons of carbon have been added to the atmosphere. Note that the total carbon content of the atmosphere is only 700 billion tons. B.Bolin,[1] one of the foremost experts on the subject, claims that man has unknowingly begun a sort of global geochemical climatic experiment that could easily get out of control. The sorcerer's apprentice performs experiments on himself without knowing how to deal with the consequences.

THE GREENHOUSE EFFECT

The surprising climatic importance of a gas like carbon dioxide, present in the atmosphere in traces (0.03 per cent), is due to a special characteristic, namely that it absorbs a part of the spectrum of radiant energy to which the other atmospheric gases are transparent. In other words, radiation emitted by the Earth which would normally pass through the atmosphere into space, is instead trapped by carbon dioxide.

The Sun emits a spectrum of radiation ranging from ultraviolet, through visible to infrared. Some of these wavelengths are partially absorbed, diffused or reflected by atmospheric gases, suspended particles and clouds. Stratospheric ozone strongly absorbs in the ultraviolet and visible regions. Water vapour absorbs mainly in the infrared. Cloud can reflect up to 70 per cent of the total incident radiation, depending on its altitude and thickness. The portion of

---

[1]　B. Bolin, 'Changes of land biota and their importance for the carbon cycle', *Science*, **196**, p. 613, 1977.

radiant energy absorbed by the atmosphere adds to the global thermal content of the atmosphere. Energy reaching the surface of the Earth is absorbed and partly reflected and returned to space; this depends on the reflective power of the surface commonly known as albedo. Absorbed energy heats the surface of the Earth.

For the surface temperature to remain stable over the period of a year, the Earth and its atmosphere must radiate into space as much energy as they absorb from the Sun. This is the only way to maintain the natural equilibrium. The Earth and atmosphere can lose energy to space only by the emission of radiation in the infrared region. The wavelength of the infrared energy returned is different from the wavelength of the radiation coming from the Sun. This is why gases like carbon dioxide and water vapour, which do not absorb the incoming radiation, absorb the outgoing infrared radiation. Increasing concentrations of these gases in the atmosphere absorb increasing

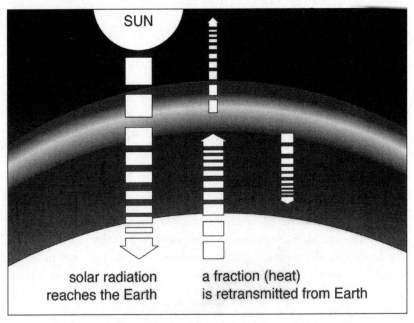

Figure 6: A simplified scheme of the greenhouse effect.

quantities of certain bands of infrared. The energy absorbed by these gases is returned to the earth as infrared radiation, increasing the total flux of radiant energy on the Earth's surface. This process is known as greenhouse effect (Figure 6). It has been estimated that the surface of the Earth would be 30-40°C colder without this additional energy flux. The surface of Mars has a temperature of 50°C below zero due to its rarefied atmosphere. In the same way Venus, where there is an abundance of carbon dioxide, has a surface temperature of +400°C.

Obviously an increase in carbon dioxide in the atmosphere leads to an increase in the amount of infrared energy absorbed and of that transmitted back to earth, and hence an increase in the Earth's surface temperature.

FORECASTS AND SCENARIOS

The surface temperature of the primordial Earth was only 30°C colder than now, even though the luminosity of the Sun is said to have been half its present value. It is widely recognised that this was due to a greenhouse effect caused by the higher concentration of carbon dioxide in the atmosphere. These concentrations were probably even higher before the evolution of organisms capable of photosynthesis and carbonate formation, which helped to reduced carbon dioxide levels. Thus, millions of years ago, much carbon dioxide was converted to sedimentary deposits: coal, oil and natural gas are fossil residues of living organisms which fixed carbon dioxide by photosynthesis. Since then the climate of the Earth has fluctuated considerably. Ice ages alternated with interglacial periods in cycles of the order of 100,000 years. Our epoch coincides with the middle of a relatively warm interglacial period of about 10,000 years. During these great climatic variations the sea level rose or fell and temperature and precipitation modified life. However these changes occurred extremely slowly (many thousands of years) and biological evolution followed at a similarly slow pace.

Shorter period climatic fluctuations due to other factors were superimposed on the long-term changes. For example the period 1550-1850 is commonly called 'the little ice age'. European glaciers advanced as never before and 1816 is remembered in New England as the year when summer didn't come. Nevertheless the variations in mean temperature of the Earth's surface, estimated from paleo-climatic data, were very small (less than one tenth of a degree per decade) and the difference between mean temperature during the 'little ice age' and today has been estimated to be about one centigrade degree. With respect to these great but slow climatic fluctuations, the impact of the industrial scale combustion of fossil fuels at an ever increasing rate and in a tiny span of time compared to biological times, may be extremely dangerous and difficult to predict. Books which deal with the climate problem from the historical point of view (otherwise excellent essays such as that of Orr Roberts and Lansford[2] or Le Roy Ladurie[3]) omit or minimize the role of carbon dioxide from the combustion of fossil fuels. The attitudes of the authors oscillate from blind faith that nature can absorb all man's damage to equally ingenuous faith that technology can repair the imbalances created. These historical studies also reveal large uncertainties of interpreta-tion. For example, there are different opinions about why the Mesa Verde villages were abandoned, especially Kayenta in the Betatakin canyon in Arizona. Some say that it was the result of a climatic varia-tion characterised by a long period of drought, others say that it was due to erosion caused by irresponsible ploughing and deforestation and the introduction of agriculture. This extraordinary ruin, and the reason for the sudden migration of the Kayenta south to the Hopi mountains, remain a mystery.

The experimental evidence of urban 'heat islands' is plain to all. Cities may be several degrees warmer than the surrounding country-side. For example in the city of St. Louis, where I worked for a time,

[2]   W. Orr Roberts and H. Lansford, *The climate mandate*, W.H. Freeman and Company, S.Francisco, 1979.
[3]   E. Le Roy Ladurie, *Times of feast, times of famine: a history of climate since the year 1000*, George Allen & Unwin Ltd, London, 1972.

one summer was very hot and 15 per cent more rain fell that year. The phenomenon was studied in depth and it was found that asphalt and concrete acted as accumulators of heat and the large quantity of fossil fuels consumed had a synergic effect. Also particulate material in the air functioned as catalyzing nuclei for the formation of water droplets in clouds.

A few climatologists may still be sceptical that human activities can heavily influence biological equilibria, however even last century several physicists had realised the problem of carbon dioxide. In 1827 L. Fourier likened the atmosphere to a glass panel. In 1861 J. Tyndall spoke of the role of carbon dioxide and the greenhouse effect. In 1896 S. Arrhenius estimated that the doubling of the concentration of carbon dioxide in the atmosphere would raise the global temperature by about 6°C.

Other scientists played down the problem, claiming that the ocean could absorb all the carbon dioxide produced by industry. This opinion was proved wrong once and for all by two studies: (1) Hans Seuss and Roger Revelle experimentally demonstrated the extreme slowness with which the deeper layers of the ocean absorb carbon dioxide. It followed that up to 80 per cent of the carbon dioxide of industrial origin builds up in the atmosphere and that even if the emissions of carbon dioxide were stopped, it would not be possible to restore pre-existing levels. (2) Precision instruments were installed on Mauna Loa (Hawaii) and at the US bases at the South Pole. These recorded carbon dioxide levels for many years and found a very steep rise in carbon dioxide in both stations.

The Mauna Loa recording is shown in Figure 7. The seasonal fluctuations depend on plant cycles. Carbon dioxide levels rose from 314 ppm in 1958 to 334 ppm in 1978. This corresponds to a nett increase of 42 billion tons of carbon in the atmosphere. Before the industrial revolution the carbon dioxide level was about 270 ppm. These measurements have been repeated all over the world, confirming the trend and revealing the highest levels in the north near the countries which consume most fossil fuels. In the next 50 years an

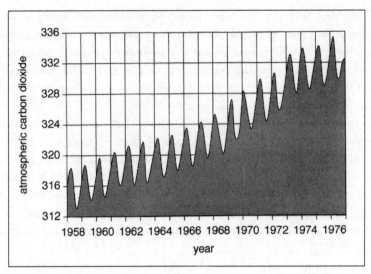

Figure 7: The Mauna Loa recording of carbon dioxide.

increase in the use of these fuels is predicted in industrial countries and a rapid rise also in the third world. The consequences may be very serious. If the trend recorded on Mauna Loa continues, the doubling of carbon dioxide levels could occur before 2030. This is infinitesimal in the biological time scale.

The increase in atmospheric carbon dioxide causes an increase in the temperature of the Earth. Estimates generally indicate that doubling of carbon dioxide levels would increase the mean temperature by about $3\,^{\circ}C$, with oscillations from $1.5\,^{\circ}C$ to $4.5\,^{\circ}C$, an increase of 7-$8\,^{\circ}C$ at the North Pole and an increased mean annual radiant energy flux of 4 watt per sq m. These effects are of the same magnitude as the temperature differences which separated the principal geological eras and can cause physical and biological variations of the same magnitude. But this time the changes induced by carbon dioxide are occurring in the space of a few decades: it is as if geological time were accelerated to force a traumatic high speed biological and physical evolution.

However there are other factors which influence the Earth's

temperature and climate: suspended materials, dust, traces of gas, water vapour, sun spots. This is why a number of theories were still given credence up to a few years ago. The main one, supported among others by Landsberg and Griffith, has been called the 'wait and see' approach. It does not exclude climatic fluctuations but claims that it is impossible to make forecasts. The second, supported by Hurd Willett, emeritus professor at MIT, is based on the 'solar-climatic' hypothesis. It predicts cooling of the Earth's surface and a sudden return to heating by about 2000-2010. Willett's forecasts are based on the study of sun spot cycles. This theory has never had many adherents. The third theory, based on the greenhouse effect, predicts an increase in temperature starting less than 20 years from now for up to 200 years. Kellogg, of the US National Center for Atmospheric Research, laid the foundations for this theory. He supplied the following data: CO_2 from 300-400 ppm by 2000, doubling of CO_2 levels by 2040. He concluded that the anthropogenic influences will dominate over the natural processes of climatic variation before the end of the century and that bigger and faster perturbations than those of the last 10,000 years will heat the Earth's surface more than has ever been recorded in the last 1000 years.

Today the introduction of refined mathematical models and an enormous mass of recent data permit us to discard the first theory, without excluding factors of uncertainty due to the influence of traces of different gases (nitrogen oxides, ozone, krypton etc) and suspended material (aerosols), the effects of which are often of the opposite sign. The three theories have correct and reasonable bases and are not regarded as being in complete contradiction. The main point is that the thermal effect of increases in carbon dioxide levels manifested very slowly and almost imperceptibly until not long ago. Only now does this effect begin to be distinguished from the background noise of the other climatic variables. This is evident from the leading studies in the field, which are examined below.

Climatologist Jim Hansen, director of the Goddard Space Center of NASA and coordinator of a group of atmospheric physicists, devised perhaps the most elaborate model for the prediction of future climate.[4] These were his data and conclusions: (a) carbon dioxide levels will reach 600 ppm in the first half of the next century, even if the use of fossil fuels increases slowly; (b) the effect of man's activities (burning of fossil fuels and deforestation) will be a clearcut heating of the Earth's surface, depending on the amount of fossil fuels consumed.

Hansen considered three situations. The first involves a rapid increase in energy (3 per cent per year). Overall heating is 4.5°C. The second situation involves slow growth (half that of the first case) and a 2.5°C increase. The third is for zero energy growth, with a temperature increase of about 1°C.

Roger Revelle, past president of the American Association for the Advancement of Science (AAAS) anticipated a 2.8°C increase in mean surface temperature due to carbon dioxide, with climatic changes which could be disastrous.[5] He added that it is reasonable to suppose that surface heating will melt snow and ice, reducing the albedo (reflectance) of the Earth. There would then be a greater absorption of solar radiation and this could lead to a further increase in temperature. Revelle analyses the model proposed by Syukuro Manabe, Richard Wetherald and Ronald Stouffer of Princeton University and comments that with the doubling of atmospheric carbon dioxide levels, the climatic changes forecast by the model are larger than any occurring since the end of the last ice age about 12,000 years ago. It is not out of the question that world temperatures rise to levels unprecedented in the history of civilized man. These high temperatures would persist for hundreds of years until the slow absorption by the oceans removes the excess carbon dioxide from the atmosphere. Manabe admits that his model is still very crude, but adds that he would be very surprised if the greenhouse effect did not eventuate. Manabe's three dimensional model predicts an increase of about 3°C with the doubling of CO_2 levels, amplified to 8-10°C at the poles.

---

[4]   J. Hansen, 'Climate impact of increasing atmospheric carbon dioxide', *Science*, **213**, p. 957, 1981.

[5]   R. Revelle, Carbon dioxide and climate, *Le Scienze*, **170**, October, 1982.

Simpler models indicate an increase of about 2°C.

In Italy Colonel Andrea Baroni of the air force meteorological service remarked that rains have become sparser but more violent, droughts more acute, every atmospheric event accentuated.

> "Spring and autumn are much shorter, in fact they hardly exist anymore. In the last 50 years the mean air temperature has increased by 2C° in the northern hemisphere. Two degrees is a lot, in such a short time. The atmospheric carbon dioxide build up is responsible for it. By 2050 this build-up will have more than doubled because of the increasing use of fossil fuels. Next century there will be peaks of carbon dioxide which will cause unprecedented climatic phenomena. It will be an extremely hot period with genuine atmospheric calamities. The heat will cause melting of part of the polar ice caps and the sea level will rise submerging the lower coastal areas."

The most detailed study of the carbon dioxide question was published in 1982 by the US National Academy of Science and the National Research Council.[6] The study follows the earlier partial study known as the Charney report, prepared by the same bodies in 1979. Dozens of scientists from the major American universities collaborated and came to the following conclusions: (a) there have been unmistakable changes in the last 20 years: the quantity of carbon dioxide in the air has increased; (b) by burning fossil fuels and converting forests to agricultural land and cities, man transfers carbon dioxide to the atmosphere; (c) the carbon dioxide increase can modify world climate, the biological systems that form the basis for life, agriculture and human society, all of which are increasingly interdependent; (d) the report discusses two studies which concluded that the effect of the carbon dioxide increase on surface temperature of the Earth would be less than estimated by most scientific groups. It

---

[6]   National Academy of Science, *Carbon dioxide and climate: a second assessment*, National Academy Press, Washington, 1982.

concludes that these studies are incomplete and invalid; (e) the temperatures observed on the surfaces of Mars, Venus and the Earth confirm the existence and magnitude of the greenhouse effect; (f) the Charney report of 1979 estimated a rise of 3°C in mean global temperature for doubling of carbon dioxide levels: no substantial revision of this conclusion is necessary. The rise will be two or three times greater at the poles, especially in the Arctic regions, with a consequent melting of ice; (g) the melting of ice will begin sooner and the first snows will fall later, shortening the spring and autumn; (h) scenarios (hypothetical climatic models in space and time) can be very useful tools and are a valid basis for decisions of immense social and economic importance.

THE DELICATE EQUILIBRIUM OF THE TROPICAL FORESTS

Twenty-two hectares of the world's tropical forest are destroyed by man every minute. This amounts to 31,000 hectares a day or 11 million hectares per year of forest lost forever. At this rate all the tropical forest will disappear in 85 years, the span of a human life, taking with it the lung required for life on Earth. It is estimated that in southeast Asia, where the demographic pressure which forces populations to clear forests for agriculture is strongest, the forest could all be gone in a few decades. The only exception is China where half a million square kilometres have been reafforested in the last thirty years.

Tropical forests are ecosystems with millions of varieties of plants and animals which maintain the equilibrium of the natural cycle of oxygen production and carbon dioxide absorption. Their disappearance would cause catastrophe, even in distant parts of the world. Another serious consequence of the conversion of tropical forests to farm land is that the humus is very quickly degraded by the sun, the soil leached and the originally fertile earth becomes desert.

Forests absorb 19-20 times more carbon dioxide per unit area than cultivated land or pasture. Their destruction causes a further input of carbon dioxide into the atmosphere, possibly of the same order of magnitude as that from fossil fuels. Woodwell's studies[7] show that the biomass, instead of continuing to be a reservoir, becomes a source of carbon dioxide comparable in size to that of fossil fuels (2-5 billion tons per year).

Bolin too agrees that the carbon dioxide increase is partly due to deforestation (wood for fuel and industrial uses) and to the conversion of forests into farm land. The limiting factor is that the increasing population requires more land to grow food, but the transition from forest to agriculture drastically decreases the amount of carbon in the world biomass, increasing atmospheric carbon dioxide. Alongside this there is the increase in erosion and deserts and climatic variations detrimental to food production. In other words agriculture and the biological equilibrium of the Earth cannot possibly survive if the present, already reduced areas of forest are not maintained. Bolin also points out that the balance of forests in developed countries, even with their programmes of reafforestation, do not weigh significantly in the carbon-atmosphere balance.

The most serious damage is occurring in the Amazon basin, Mexico and Central America. In the latter two places, two thirds of the forests have already been destroyed and the process should be complete in less than twenty years. The tropical rain forests are transformed into pastures for US beef production. James Nations and Daniel Komer, two American ecologists, are the authors of a damning article entitled 'Rainforests and the hamburger society'[8] from which some alarming points emerge: (a) in Salvador the forests have already been completely destroyed; (b) in the other countries of Central America deforestation reaches as much as 1000 sq Km per year (Table 4); (c) in Central America 90 per cent of the population increment will settle in areas now covered by tropical forests; (d) generous loans are offered by governments and banks for the transformation of forest into farmland. After two crops of maize, rice or manioc, the delicate

---

[7]   G.M. Woodwell, *Science*, **199**, p.141, 1978.

[8]   J. Nations and D. Komer, 'Rainforest and the hamburger society', *Environment*, **3**, p. 12, 1983.

TABLE 4: DEFORESTATION IN CENTRAL AMERICA		
Country	**Existing forest (1982)** (sq kilometers)	**Area lost/year** (sq kilometers)
Nicaragua	28,000	1,000
Guatemala	26,000	600
Panama	22,000	500
Honduras	20,000	700
Costa Rica	16,000	600
Belize	9,800	32
Mexico	8,000	600
El Salvador	0	0
Total	130,000	4,132

soil, previously protected by the rainforest, can no longer support crops. It is then turned into pasture for cattle which are sold to the US for hamburgers. This continues for ten years, by which time the land has become desert; (e) more than 90 per cent of the meat produced is exported to the US. *The Americans eat hamburgers without realising that they are eating toucans, tapirs and rainforest.* This exploitation began after the last world war at the initiative of the American food industries.

Prior to the 'hamburger deal', the forests were mainly exploited for mahogany and tropical cedar. The FAO reports that 93 per cent of the land controlled by seven per cent of owners. In Guatemala 2.2 per cent of the population owns 70 per cent of the agricultural land (coffee, bananas and pasture). Anastasio Somoza owned 23 per cent of the cultivable land in Nicaragua.

It is interesting that after only five years the annual yield per hectare in meat is only ten Kg, whereas the Maya Lacandones, with their traditional agricultural methods, produced 6000 Kg of maize per hectare per year and 5000 Kg of vegetables. Such yields could be sustained for seven consecutive years, after which the land was abandoned for ten years, alternating cycles of agriculture and forest on the same land. This is possible with a type of mixed agriculture (cocoa, avocado, papaya, rubber and citrus trees) which conserves the humus of the rainforest and uses it as a renewable resource.

It is worth recalling that during the war in Vietnam, about 100,000 sq Km of forest was lost through bombardment and chemical defoliation.

OZONE

The ozone molecule consists of three atoms of oxygen joined in an unusual way. Ninety seven percent of atmospheric ozone is found in the stratosphere, but its concentration is very low. It absorbs most of the ultraviolet spectrum, preventing this radiation from reaching the Earth. The longer ultraviolet wavelengths (so called UV-B from 29,000-32,000 nm) are only partly absorbed by ozone. If the ozone is reduced by 10 per cent, the UV-B radiation reaching the Earth increases by 20 per cent. This type of ultraviolet is responsible for the attractive suntan, but in slightly higher doses can cause a variety of skin cancers including melanoma. It probably also causes damage to natural ecosystems: mutations, partial inhibition of photosynthesis, reduced plant growth.

Stratospheric ozone, which protects us, may be reduced by the microbial transformation of nitrogenous fertilizers and oxides of nitrogen produced by supersonic jets. However these two effects seem minimal. To the contrary, the effects of chlorofluoro-methanes (CFM, known as Freon) are severe. These inert gases are used as propellants in spray cans but mostly as refrigerants in air conditioning and refrigeration plants. When released they migrate, unmodified, to the stratosphere where they are broken down by ultraviolet radiation, releasing chlorine and triggering a chain reaction which destroys ozone.

Predictions based on Freon emission in 1973 suggest a 14 per cent decrease in ozone levels and a 28 per cent increase in ultraviolet radiation on the Earth, with the same increase in skin cancers. The use of Freon was prohibited in the US in 1978, and in Sweden in 1979.

The problem is extremely serious in view of the low concentrations and high instability of ozone in the presence of chlorofluoromethanes, also because once the destruction of ozone is underway, nothing can be done about it. A side effect of CFMs is to enhance the greenhouse effect. Ramanathan claims that a few parts per million of CFMs can lead to a significant rise in temperature.

ACID RAIN

Many and complex effects are expected from the increasing consumption of fossil fuels. The planetary scale of these effects is certain to upset planetary equilibria (energy balance, climate, agriculture etc.) with serious consequences for man. The main chemical products of fossil fuel combustion responsible for environmental damage are carbon dioxide with its greenhouse effect, and oxides of sulphur and nitrogen with the resulting acid rain. Most scientists consider that the three environmental time bombs are the greenhouse effect, acid rain and contamination by toxic substances.

Coal, oil and the exhausts of cars are sources of sulphur and nitrogen oxides. A thermoelectric power station can release up to 400,000 tons per year of sulphur dioxide (SO_2). In 1980 the USA discharged 26 million tons of sulphur dioxide and 22 million tons of nitrogen oxides into the atmosphere; Italy produces about two million tons of sulphur dioxide per year. USA, Canada and Europe produce more than 100 million tons of sulphur dioxide in a year.

These oxides circulate, are taken up in clouds and transformed by water vapour into sulphuric and nitric acid; the resulting rain can fall hundreds of kilometres from the emission source.

Aquatic ecosystems such as lakes and rivers are the most sensitive to acid rain. Many US and Canadian lakes have suffered irreversible damage. In Sweden 4,000 lakes are without fish and 14,000 have acidified considerably. The responsibility for this lies with the northwest American and northern European industrial belts, in the case of Scandinavia. In general the whole industrialised northern hemisphere is affected, except for a few geographical areas which tolerate acid rain because of the neutralizing effect of the soil.

Figure 8 shows the effect of acid rain on certain aquatic animals. The scale shows a range of acidity values in terms of pH;[9] pH 7 is neutral (e.g. distilled water); lower values indicate acidity and higher values alkalinity (basicity). Note that the acidity of rain tends to increase and at pH 5-4.5 fish can no longer live. Molluscs die at pH 6; only certain water beetles can survive beyond this. Rain acidity is increasing: in 1979 the snow that fell on Montana had a pH of 2.6; in Scotland they breathe pH 2.5 fog; in Milan in January 1983 pH 3.6 rain was recorded. Further harm is perpetrated upon animals, plants and man due to the fact that acid rain mobilizes certain very dangerous heavy metals: mercury, lead, cadmium, nickel and plutonium.

Acid rain modifies soil chemistry, depriving plants of nourishment. Calcium and potassium (essential for plants) are leached out. The pH has changed by one and a half units in a century. Research has revealed a 50 per cent loss in crops and increased vulnerability of leaves to disease. Beate Weber, a West German representative at the European Parliament, claims that 560,000 hectares of forest have been damaged in his country.

Other effects of acid rain include the corrosion of buildings, steel bridges and monuments; the list of damaged works includes the cariatids of the Acropolis of Athens, Egyptian temples, the horses of San Marco in Venice, the churches of Krakow, the cathedral of Koln. In Sweden, damage due to corrosion amounts to 800 million dollars and the Epidemiological Division of Brookhaven National Laboratory estimates that there are 7,500-120,000 deaths per year in the US due to acid derivatives of sulphur from fossil fuels. They also ascribe many respiratory diseases (asthma, emphysema, chronic bronchitis) to acid rain.

Solutions exist, if we are prepared to accept them: (a) reduce energy consumption; (b) use clean or renewable fuels (methane, sun, bio-

---

[9] Hydrogen potential or pH is the conventional measure of acidity (or alkalinity) of aqueous solutions as a function of the concentration of hydrogen ions.

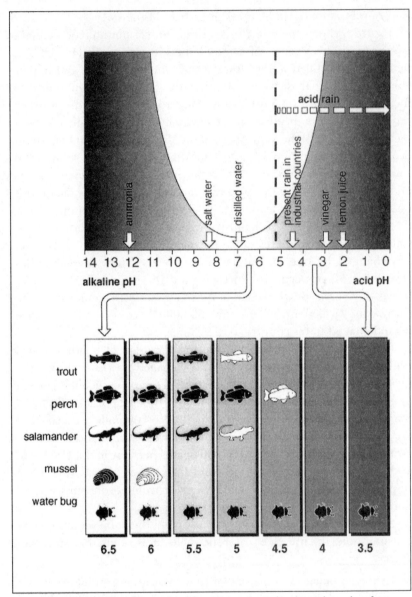

Figure 8: The effect of acid rain on certain aquatic animals.

mass); (c) use appropriate technologies to scrub emissions of sulphur dioxide. The realisation of these provisions would create new possibilities of employment. The cost of reducing emissions would be less than the hidden costs that acid rain is currently debiting to society.

CLIMATE AND SOCIETY

During a recent space trip, an astronaut commented that the atmosphere surrounding the Earth was so polluted that it had lost its limpidity and transparence.

Complex problems like stratospheric ozone, acid rain and the greenhouse effect require immediate measures. The stakes are enormous: irreversible modifications to climate, damage to biological equilibria, reduced food production and serious social repercussions.

Figure 9 shows the trend of mean global temperature over the millennia. A period of natural cooling was the expected trend, but the

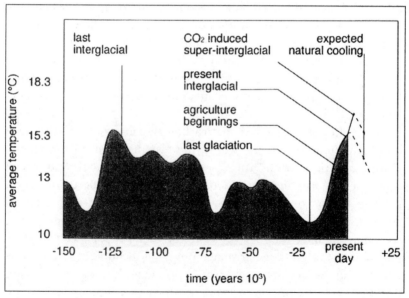

Figure 9: The trend of mean global temperature over the millennia.

greenhouse effect is causing a super-interglacial period and the highest temperatures in the last million years are predicted next century. This means melting ice caps, rising sea level, flooding of coastal cities and farmlands (the gondolas in New York of *Environment*) and the opening of the legendary 'northwest passage'. Carbon dioxide will cause an increase in water vapour and the humidity of the atmosphere will contribute to the greenhouse effect. The overall result will be a further increase in temperature and the disappearance of the spring and autumn: great summer droughts and desertification of the semi-arid areas of the world including southern Europe. A trend of this kind is already being recorded in Spain and southern Italy.

The greatest harm will be suffered by areas close to deserts and third world countries with already precarious agriculture. In 1983 drought struck Ghana and the Ivory Coast, and the hot dusty wind (Armattano) from the desert aggravated the situation. The desertification of Africa associated with carbon dioxide build-up has already begun. If the process continues, the arable lands of the Earth will be displaced towards the poles. The south of the northern hemisphere, tropical countries and the third world will pay for the industrial activities and crazy energy consumption of the developed countries. Carbon dioxide is turning our planet into a desert hot-house, placing a mortgage on food production.

Roger Revelle remarked that carbon dioxide could be the most important substance in the world; however he went on to imagine absurd solutions with unforeseeable environmental and social consequences like moving seaboard cities, large scale construction work in the major river basins, to move water from one valley to another, new agricultural products created by genetic engineering etc. Again we have the spectre of the sorcerer's apprentice who does not want to forego the charismatic power of technology, even if it is impotent in the face of planetary scale problems. Without considering the repercussions on labour and the habits of human society, we cannot ignore that most animal and plant species would require millions of years to adapt genetically to such large and sudden changes in temperature. These changes will cause the extinction of many forms of life.

If human society wants to decide the course of its future and the life of this indispensable planet, it must adopt the main conclusion of the US National Academy of Science: "the main limiting factor of energy production from fossil fuels in the coming centuries could be the climatic effects of the liberation of carbon dioxide." This means a transition from fossil fuels to clean renewable energy sources and a limit to growth of energy consumption.

CHAPTER 8
THE GREEN SCENE:
AGRICULTURE IN TERMS OF ENERGY

The conservation of natural resources is necessary for human survival and for sustainable development. For the future of the planet it is essential that three things be maintained: ecological processes, genetic diversity and the stability of the systems upon which natural equilibria are based. Agriculture has a prime role in such a strategy. Aggression to the environment, waste of resources, deforestation, pollution, impoverishment of arable land and population growth are the adversary's pawns in a diabolical chess game which envisages the death of winner and loser in the case of defeat.

The population increase is what causes most concern at the present time. Zero population growth in the industrialised countries does not solve the problem of the disproportionate consumption of resources. A Swiss consumes as much as 40 Somalians. In order to avoid dramatic death from hunger in the third and fourth world countries there is no other solution but to scale down our waste of resources and food, and institute a strict strategy of environmental protection.

Energy is the key element in both the economic-productive and biological contexts.

As man advanced from primitive farmer to his present status, energy consumption has grown remarkably because the item 'food' has come to absorb much less energy than housing, commerce, transport and industrial activity. Modern man spends much more energy than is strictly necessary.

In a traditional small agricultural community forming part of, or at least not extraneous to, the natural environment, solar energy is the source of sufficient agricultural yield for domestic and wild animals

127

and man. The residues of this activity (which I do not want to call wastes) return to the earth where they are transformed by micro-organisms into organic material suitable for plant life. Hence there is no waste of primary materials and the energy flow is minimal and used with maximum efficiency.

On the contrary, in industrial societies the cycle is not closed and the energy input for production, transformation, transport and distribution of food, and also for waste treatment is very high. This is due to the separation between the sites of production and utilization. Nutrients are taken from the soil and brought to the towns where they become wastes, and more energy is needed to dispose of them. The impoverished land is again fertilized with artificial substances produced by consuming fossil resources, in an endless chain of waste, fed and encouraged by the cheap availability of fossil fuels.

It is obvious that a model based on non-renewable polluting resources cannot be sustained for long, but this does not mean that we must return to the peasant's way of life. The solution is to plan a model of life based on renewable resources with maximum efficiency in the use of resources and an end to waste. In this way many of the advantages that we currently enjoy could be maintained and extend-ed to the rest of the world population, thanks to scientific knowledge and appropriate technology. It is obvious that agriculture is the hub of such a model of life.

AGRICULTURE IS CAPTURING ENERGY FROM THE SUN

Agriculture is a human activity which brings energy into our system, the Earth. Over the millennia man has not created energy (the first law of thermodynamics states that energy cannot be created or destroyed) but has enriched the Earth with energy from the Sun which would otherwise not have been used.

The intuition of certain American ecologists (Odum, Commoner and others) is helpful here. The purpose of agriculture is to produce food (meat, milk, vegetables, cereals) and materials (wool, linen, cotton, hemp, wood), but from the point of view of energy, *agriculture is capturing energy from the Sun.*

The analysis of ecosystems in terms of energy has gradually been extended to the simplest agricultural systems. Most of the energy that the land receives from the Sun is dissipated, for example by reflection, accumulation in the soil and water and in causing water to evaporate. These functions are all essential for the organisation of life on Earth: climatic regulation, the maintenance of chemical cycles, the flow of living and non-living materials (transport of seeds and nutrients).

Only a small percentage (about one per cent) of the solar radiation incident upon fertile land is fixed by chlorophyll photosynthesis in plants in the form of high energy organic molecules. The plant transforms this energy by biochemical processes (respiration) into organic compounds and work.

When examined in terms of energy flow the food chain becomes intelligible: energy is progressively degraded in the successive phases of the chain (plant producers, animal consumers, microorganism decomposers), returning to nature the elementary substances necessary to rebuild, with the help of solar energy, the molecules of living cells.

The entire organisation of living creatures in mature ecosystems minimizes the dissipation of the energy fixed by plants, using it completely for its complex regulation mechanisms. This is made possible by great energy reservoirs (biomass) and the diversification of living species. However this stability of natural ecosystems means that the final energy yield is zero, except for a relatively small quantity of biomass which is buried to form fossil fuels for the future.

Traditional agriculture is a very simplified ecosystem in which the biochemical processes minimize the energy dissipated in the phases of the food chain and accumulate large quantities of energy in plants which can be eaten by man or animals. By contrast in industrialised agriculture, the regulation or alteration of the ecosystem for productive purposes is achieved by consuming fossil fuels, i.e. by increasing energy input.

As a result of the energy choices of the last 50 years, agriculture not only captures less energy but also contributes to the irreversible decadence of energy sources, transferring its weight from the positive to the negative side of the scales. In other words the energy input of agriculture is increasing because of the often irrational and thermodynamically absurd use of petroleum products (fertilizer, pesticides etc.), electricity and fuel. The use of biomass for energy production is far from being developed, not only because of scientific and political problems but also due to the arbitrary pattern of world agriculture which does not meet real food and energy needs, especially in the third world.

We need to adopt a new form of energy-producing agriculture which makes use of renewable sources and is no longer subordinate to industrial choices. Genuine energy farms are a real possibility. They would constitute an alternative to fossil fuels and nuclear energy. This transformation requires that agricultural operations be evaluated by energy criteria.

A significant increase in efficiency of the system can be obtained by increasing and qualifying output: new unconventional sources of food, coproducts for energy or fertilizer, integration into the natural and human environment, transforming polluting effluent into savings in energy and materials. In a strategy of modification of the overall structure of agriculture of this kind, the experience of 'organic' farmers and all the scientifically-based experiments leading to maximum use of natural methods with minimum input of non-renewable energy, are of value as pilot studies.

Renewable energy sources such as the sun, biogas, biomass, wind, minihydro and geothermal, have tremendous potential in agriculture. Agriculture offers the possibility of a market which is already partly suited to the use of renewable resources, because of the distribution of users, the seasonal nature of certain operations which coincide with maximum solar availability and the need for low temperature sources. Research and development of new renewable sources will be of great assistance to emarginated zones. For example photovoltaic energy can be used in areas where connection to the national network would be too costly.

An attempt to tackle the energy problem in agriculture and disengage it as much as possible from the use of petroleum products does not mean merely to find applications for solar energy or cultivate biomass, but to transform the whole agricultural system in an economic, scientific and cultural effort which will restore quality to labour and farm life. Even if there is no significant short term contribution by agriculture to solving the energy problem, any policy of environmental restoration and safeguarding of resources must centre around a relaunching of agriculture and the utilization of areas of low urbanization. Only in this sense can a 'return to the soil' be a practicable and indispensable path for the future.

An adequate exploitation of photosynthesis could help to supply man with enough energy and food before the depletion of fossil fuels, i.e. fuels produced by photosynthesis millions of years ago. In researching this and other alternatives, it should be remembered that even remedies cost energy. It is necessary to spend a large slice of our energy resources to equip ourselves to do without fossil fuels. This task should not be delayed.

Every year the Sun sends 5.6×10^{24} joule of energy to Earth. Part of this is reflected by the upper atmosphere or absorbed by the atmosphere itself. Another part is absorbed by oceans and land. Only a small part is absorbed directly by photosynthesizing organisms, the only ones capable of converting it to chemical energy. The process of photosynthesis consists of forming starch molecules from carbon dioxide and water using solar energy:

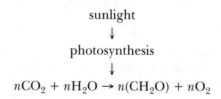

$$\text{sunlight}$$
$$\downarrow$$
$$\text{photosynthesis}$$
$$\downarrow$$
$$n\text{CO}_2 + n\text{H}_2\text{O} \rightarrow n(\text{CH}_2\text{O}) + n\text{O}_2$$

This is a simplified scheme, because plants also produce proteic compounds from nitrates and this too, requires a little solar energy.

Simple calculation shows that only a tiny part of the solar energy is converted by photosynthesis. The most plausible hypothesis poses a theoretical limit of about six per cent, however there are still doubts about certain interpretations of the photosynthetic process.

Spontaneous photosynthesis in uncultivated areas in favourable climatic conditions is very efficient, because the land is covered year round with vegetation adapted to maximum exploitation of solar energy. By contrast, domestic cultivation involves less intense periods and crops which, although necessary to man, are less efficient. Hence it is not surprising that the overall photosynthesis on the Earth only has an efficiency of about 0.15 per cent. Only in areas of particularly intense modern agriculture is it possible to reach one per cent, and four per cent with certain crops (sugarcane, sweet sorghum) but only with massive external energy input.

World production of organic material (biomass) is about 2×10^{11} tons per year. This has an energy content of about 3×10^{21} joules. Man's crops account for only five per cent of this biomass. Since the yearly energy consumption is 3×10^{20} joules, this constitutes one tenth of the photosynthetic product.

Hence the task is to find a way to exploit this enormous renewable energy source without depriving man of food, without using fossil fuels and without devastating the ecosystem with excessive simplifications which can only be short-lived. In a word, without expecting nature to sustain the distorted and suicidal lifestyle which fossil fuels have supported up till now.

THE BIOMASS, THE FOOD-ENERGY DILEMMA AND FUEL ALCOHOL

The biomass is organic material having an energy content, and consists of all land and aquatic plants, their residues and residues of transformation by animals (manure) and by technological processes. Industrial refuse (especially from the food industry), town wastes, agricultural residues (animal and vegetable), ligno-cellulose crops, sugar and starch crops, water culture (microalgae and macrophytes) all make up the biomass. In fact the biomass is renewable solar energy distributed everywhere and technologically easy to use.

Many biomass technologies are economical and feasible from an energy point of view even on a small scale, and favour the decentralization of energy production. Energy can be extracted from the biomass with greater safety and with fewer problems of pollution and environmental imbalance, than from conventional sources.

An important aspect of biomass energy production is that the carbon dioxide is recycled: the growing biomass uses the carbon dioxide produced by the processing of the previous crop (Figure 10). By contrast, when fossil fuels are burned the resulting carbon dioxide accumulates in the atmosphere. Hence the substitution of fossil fuels with renewable energies should be one of the first steps taken to slow the catastrophic effects on climate, agriculture and food production of carbon dioxide build-up.

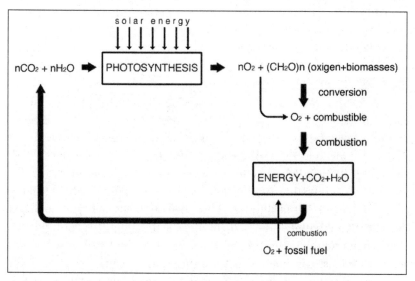

Figure 10: The carbon dioxide cycle (biomass and fossil fuels).

Biomass can be converted to fuel by chemical or biological processes. From biological process three types of fuel can be obtained. (1) Alcohol; (2) Biogas (54-70 per cent methane, 27-40 per cent CO_2, nitrogen, hydrogen, carbon monoxide, oxygen, hydrogen sulphide) is the final product of a microbiological process in which organic material is decomposed anaerobically, i.e. in the absence of oxygen. Manure is the most usual material used for the production of biogas; (3) Hydrogen is an energy resource of great potential. Research is underway into the production of hydrogen by photosynthesis from green and blue-green algae and photosynthetic bacteria.

To emphasise the absurdity of reducing the energy problem to the production of electricity, Amory Lovins likens the power station debate to choosing the type of brandy to fuel one's car. Electricity is required for only eight per cent of all our energy needs, says Lovins. The other 92 per cent is substantially heat and fuel for transport for which electricity would be uneconomical and thermodynamically incorrect. Paradoxically, this ironical simile comes close to the truth.

Ethyl alcohol (ethanol) seems to be a promising alternative to petrol. This is a true form of solar energy stored in different plants by photosynthesis. As long ago as 1932, an Alfa Romeo P2 won the Mille Miglia on a mixture containing 25 per cent ethanol. Alcohol can be mixed with petrol (gasohol) up to 10-20 per cent without substantial changes to the motor. Alcohol increases the octane rating of petrol, so that refining costs can be cut. Cars that can function on 95 per cent alcohol have recently been built. Ethyl alcohol has a lower energy density than petrol, but given its high ideal compression ratio, the power developed per litre is higher than for petrol, giving more Km per calorie.

Another interesting combustible alcohol is butandiol, which can be produced from hemicellulose. This dialcohol has a higher energy density than ethanol and can form more stable mixtures with petrol.

Biomass suitable for the production of alcohol can come from sugar crops (sugar beet, sweet sorghum, sugarcane), starch crops (cereals, potatoes, topinambur, manioc) or ligno-cellulose crops (forestry residues, poplar, eucalyptus etc.). Billions of micro-organisms can be used to create energy in the fermentation process. The micro-organisms traditionally used in industrial alcoholic fermentation are the yeasts, especially *Saccharomyces cerevisiae*. Even certain bacteria can bring about this process, for example *Zymomonas mobilis* and certain species of *Clostridium* (*thermosaccharolyticum* and others) which are heat-stable and can therefore be used in new high temperature fermentation techniques like continuous low pressure fermentation. In this process the ethanol is extracted continuously by virtue of the high-temperature-low-pressure regime, so that it never reaches concentrations high enough to inhibit microbe metabolism. The fermentation, which traditionally proceeds in two phases, can thus go on without interruption. Tricks like recycling and immobilization of cells, and dilution of ethanol by dialysis have been tested in laboratories and pilot plants.

The environmental advantages of using alcohol for fuel can be considerable. Atmospheric pollution from lead can be eliminated because tetraethyl lead (added to petrol to raise the octane rating) is no longer necessary. The emission of sulphur compounds and particles is reduced to zero. Nitrogen oxides and carbon dioxide are halved. However it is still necessary to control the emission of products of incomplete combustion (aldehydes and alcohol itself).

To produce alcohol to replace conventional liquid fuels is an attractive enough idea for countries which import petroleum. Already many countries have programmes of alcohol production in various stages of operation. By far the most ambitious and advanced is the Brazilian Proalcohol Project: in 1985 10.7 billion litres of ethanol were used. This is about 40 per cent of the country's fuel consumption.

However the production of alcohol may be integrated into the agricultural situation of a country in different ways: the social and economic effects are different as are the effects on the environment and food production. In Brazil these effects are all negative and the outlook is very bad. Proalcohol is based on a model of waste, indiscriminate exploitation of nature, massive investments, centralization of energy and non-integrated technology. Forty percent of its enormous plants have a capacity of more than 50 million litres per year. Vast areas are cultivated with sugarcane which is transported long distances to the distilleries. Energy inputs of this kind weigh on the energy balance.

These agro-industrial mega-units introduce sophisticated modern technology into the countryside. In order to be run efficiently they require maximum exploitation of the soil and expansion of the sugar-cane monoculture. Sugarcane is replacing food crops: in 1985 13 million hectares of fertile land were dedicated to sugarcane instead of producing five million tons of maize or 4.5 of rice or 1.5 of beans.

At the social level, the benefits of increased employment opportunities in the rural sector are outweighed by the expropriation of small and medium landowners. In an environment where the Amazon rainforests are indiscriminately destroyed, upsetting the delicate ecological equilibrium and the life of its autochthonous inhabitants (the Indios) to create pastures or to feed gigantic paper factories, the consequences of the expansion of sugarcane can be disastrous.

Finally, the distillation residues are not used. This is certainly a greater problem for giant plants because of the quantity produced. The residues are dumped into the sea or rivers near the distillery, causing serious pollution problems. In 1981 the volume of residues dumped was equivalent to the wastes of a city of 160 million people.

The employment balance is negative: in the state of Maranhao there were 100,000 new jobs and 400,000 more unemployed farm workers. Even the food-energy dilemma is serious in Brazil: in the state of Sao Paulo, the major alcohol producer, the area cultivated with sugar cane increased by 50 per cent between 1974 and 1979, while the area of rice, millet and potatoes decreased by 35 per cent, 18 per cent and 23 per cent respectively.

In Asia, according to Roger Revelle, the food-energy dilemma is complicated by the high population density on the cultivated land. The production of fuel from biomass could lead to an increase in the price of food with negative effects on the poor.

All this demonstrates the need to examine every situation in detail, even in the same country or region, before the best solution can be identified. The best solution considers the agricultural, industrial and energy aspects and respects the characteristics of the particular area.

It can be seen that plants of small to medium capacity (0.5-4 million litres per year) have certain advantages: (1) maximum energy efficiency; (2) rapid construction (4-5 months); (3) small land requirements and thus small energy input for transporting the raw material to the distillery; (4) a small volume of residues which can be used as fodder.

The major problem of alcohol production (apart from the fact that it subtracts from food production) is not technological but concerns land availability. Most cultivable land is already used for food (1.5×10^9 hectares = 11 per cent) or for fodder (3×10^9 ha = 22 per cent); the rest of the land is forested (4.1×10^9 ha = 30 per cent) or arid and uncultivable (5×10^9 ha = 37 per cent).

Hence there are a number of possibilities: (1) increase the area of land for energy crops, to the detriment of food production and forests (as in Brazil); (2) give priority to the use of waste biomass (urban and agroindustrial wastes) and then decide whether to dedicate some agricultural or marginal land to energy crops (it is thought it would be possible to do this in Italy, France and other European countries); (3) plan an agro-industrial system capable of fulfilling food requirements and at the same time supplying sufficient biomass to produce fuel for the system itself, so as to at least partly substitute fossil fuels. This was proposed to the US government by the Energy Subcommission, headed at the time by Edward Kennedy. It was based on the findings of the Centre for the Biology of Natural Systems (CBNS) in St. Louis and the Department of Biochemistry and Biophysics of the University of Pennsylvania.[1]

Apart from urban, agricultural and industrial wastes and the utilization of marginal areas for arboriculture (fast growing species such as poplars and eucalyptus), this project envisages intervention in the fodder sector. When alcohol is produced, only the carbohydrates

---

[1] Joint Economic Committee, US Congress *Farm and forest produced alcohol: the key to liquid fuel independence*, USGPO, Washington, 1980.

ferment leaving nitrogen content relatively constant. The nitrogenous fraction has a higher nutritive value than the starting material, probably because of the increased digestibility of its components and the presence of yeasts developed by fermentation.

By changing the present system of rotation of certain prevalently fodder crops and introducing a particular rotation of maize, sugar beet and hay studied by the CBNS, it is possible to increase the carbon content of the biomass produced from a carbon/nitrogen ratio of about 20:1 up to 30:1. The extra carbon can be converted to ethanol and the distillation residues, the calorie and protein content of which are the same as those of traditional fodder, can be used as animal feed. This allows considerable saving if realised on a medium scale. From this different use of land dedicated to cattle feed, the USA could produce 175 billion litres of ethanol per year, to which would be added another 15 billion litres per year from surplus grain and food industry residues.

Other contributions could come from lignocellulose once the difficulty of converting hemicellulose has been solved. An estimated 150 billion litres per year of ethanol could be obtained from cellulose and 230 billion litres per year of ethanol equivalent (n-butandiol) from hemicellulose. On presenting this project, Senator McGovern wrote that the United States was capable of completely substituting petroleum based transport fuels with alcohol produced by renewable biomasses without affecting food and fibre production.

Obviously this project cannot simply be transposed into the European situation because of differences in crops, climate, final destination and methods of production. The crops and areas suitable for such utilization need to be identified with particular care. An excellent example is the study on topinambur performed by the École Supérieure de Formation Agricole d'Angers in France.

An interesting EEC report on the subject of alcohol was compiled by Marie-Angèle Farget and Dirk Ahner of the Direction Générale de l'Agriculture. It considered agricultural and forestry wastes and residues (e.g. straw, twigs etc.) and plants suitable for the distillation of alcohol. Ahner and Farget analysed the energy biomass in terms of energy flows. They attempted to establish a strategy which could help solve surplus and energy problems within the framework of European agricultural policy.

As regards the energy balance of alcohol production in terms of net energy produced, the higher thermodynamic yield of alcohol as a fuel and the increase in octane rating (energy savings in the refinery) must be considered. It follows that the energy balance is always positive. Small units allow a further energy gain as a result of greater efficiency and transport savings.

Until recently, criticism of the use of ethanol as a fuel was based on the low efficiency of its energy balance. The CBNS of Washington University replies that these criticisms commit at least three errors: (a) they are based on obsolete fermentation and distillation processes; (b) they do not value or undervalue the energy content of the residue used as fodder; (c) they do not consider the contribution of ethanol to the octane rating and its fuel properties in terms of kilometre per calorie.

As for any other fuel, a project for alcohol production from biomass on a regional or national scale must first consider the availability and quality of primary materials. As we saw earlier, there are many different ways of analyzing and approaching this problem, each of which gives different results not only in terms of energy but especially in terms of agricultural development and environmental impact. Once this is clear it is necessary to define the characteristics of the operational units (farms, conversion plants etc.) on which to base the project with a view to the desired production and environmental results. The energy balance of a process is also linked to plant size and its territorial integration. For small and medium plants (0.5-4 million litres per year) which are also more efficient, the energy consumed (inevitably liquid fuel) and the economic cost of transporting the primary material to the plant is less than for big plants. These need to be fed by a larger area, proportional to their size. Small to medium plants are also more easily integrated into the existing farming structures because the volume of residues left after extraction of alcohol can be completely utilised for fodder in the zone of production. The combination of alcohol plant and zootechnics (Figure 11) provides two big advantages in terms of economy and energy: (1) the optimum use of the residues (any other use would amount to waste of high quality material); (2) given the proximity of the two units, the material can be transported without processing (e.g. drying alone would require 15-30 per cent of the energy input of the entire process).

The next step to improve food-energy production and to close the natural cycle of products and residues of these farms is the anaerobic digester. Here animal and agricultural residues which could not, or could not economically be converted to alcohol, come together.

The products of anaerobic digestion are biogas (methane + carbon dioxide) and high quality organic fertilizer. Biogas can be used as it is or efficiently converted into electricity and hot water, which can be used in many parts of our system, for example the still, the fermenter, the digester itself and the stable.

The use of the residual sludge of the digester for fertilizer is a very important part of the cycle. It enables the organic material and minerals necessary for the maintenance of its physical and biological structure to be returned to the cultivated land. Depending on the size of the digester, it is possible to meet, partially or completely, the fertilizer requirements of the farm. This means economic and energy savings on organic and chemical input and environmental-energy advantages from the use of organic rather that chemical fertilizers.

There are only a few valid studies on this last point. In one,[2] 14 conventional farms are compared with 14 'organic' farms in the US Midwest. It was found that the organic farms, with a slightly lower production per unit area (10 per cent for maize, 5 per cent for soy), had the same economic yield per unit area as conventional farms because of lower costs, and above all a much lower use of fossil fuels (about 40 per cent due principally to the saving on nitrogenous fertilizer).

---

[2]   W. Lockeretz, G. Shearer and D. Kohl, 'Organic Farming in the Corn Belt', *Science*, **211**, p. 540, 1981.

These are the basic components of an energy- and food-producing farm, mainly based on livestock (Figure 11). The energy produced is in the form of alcohol, electricity, heat for internal and external use.

The primary material is supplied by agricultural residues such as those of maize, beets, fodder and pastures with certain rotations. The scheme refers to a single farm but can be extended to larger areas: the products and residues of a number of farms can be processed in common plants, the size of which needs to be decided case by case.

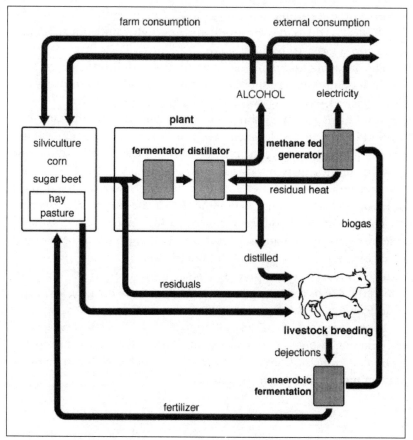

Figure 11: The combination of alcohol plant and zootechnics.

This model of an integrated farm was designed for a moderately industrialised form of agriculture but can be adapted to the level of underdeveloped countries. For example in Mexico integrated systems which could provide electricity for small, as yet unpowered villages are being studied. If alcohol production were also included, self-sufficiency in energy could be achieved, providing an incentive for renewed agricultural activity in these areas. Another possibility could be the recycling of carbon dioxide, by-product of alcoholic fermentation and anaerobic digestion (40 per cent of biogas) with cultures of autotrophic micro-organisms (photosynthetic microalgae and bacteria) which in turn supply further highly nutritional biomass to 'export' or return to the system.

The idea of modifying the agriculture and livestock system to obtain food and energy therefore seems to be a step towards solving the food and energy crisis, and reducing the use of fossil fuels. If properly managed, the plant world can supply the foundations for a form of biological development which uses sunlight as energy source and air (nitrogen, oxygen, carbon dioxide, hydrogen), water and small quantities of minerals as material resources. In this way the physical and chemical development model, based on fossil fuels, uranium and large quantities of minerals, can be relegated to the past.

The use of decentralised resources, which need to be examined case by case, does not automatically lead to democracy or better quality of life, but it certainly lays the foundations. The same is true for reducing environmental pollution: since it was created by centralization and waste of resources it can be countered by changing social and production models.

We have seen that there is still time to make this conversion without trauma, using the remaining fossil resources and appropriate technology which we already have, but every delay will make this conversion more painful and less effective.

CHAPTER 9
THE CHRONOVISOR:
ECO-NOMY OR ECO-LOGY?

I would not liken plastics to the devil and wood to holy water. Plastics have greatly improved our quality of life. We only have to think of their use in medicine and hospital hygiene. It is not so easy, instead, to understand the substitution of natural materials like cotton, wool, linen and wood with synthetics (non-renewable petroleum products) for short term profit. In my home we have been using only wood, terracotta, straw, natural fibres, iron etc. for many years. A young doctor brought this to my attention one day (our choice had been natural, not a question of principle) with the comment that I lived in an 'ecological house'!

The problem is not whether to say yes or no to a hi-fi made mostly of plastic. The problem today is to say no to the chronovisor market and to the consumeristic use of the chronovisor. It is quite possible to live without a chronovisor (something which does not exist, since I just invented this word) until industrial production invents it and the mass-media impose it upon our consumerism, passing it off as the latest drug which we cannot do without. Just as I now would not know how to do without my hi-fi. The problem is to prevent science and technology from following the laws of the market. It should be science (and its discoveries in harmony with biological equilibrium and ethical growth) that dictates the law to technology and the market.

The following comments of Orio Giarini[1] help us to understand the profound difference between the terms of science and technology in the modern age. Science, within the limits of its uncertainties and non-neutrality, tries to increase our understanding; technology tries

---

[1]   O. Giarini, *Dialogo sulla ricchezza e il benessere*, EST Mondadori, Milan, 1981. This book is one of the famous reports to the Club of Rome.

to make use of it. Science has improved our knowledge of matter; technology takes practical advantage of it. As such, technology is completely conditioned by the social, economic and institutional environments. The supremacy of science over technology in terms of priority of choice and intellectual analysis, is undisputed. A purely technocratic state would be an imbecilic state.

Technology is subject to the law of diminishing returns. Giarini's article analyses this law in detail in the different sectors of production and finds that technological returns began to decrease in the early seventies. Some of the examples he cites to illustrate this law show that the timing of technological development does not easily coincide with that of supply and demand. Other causes include the increasing production of wastes, specialisation, increasing vulnerability of the economic system (in terms of risk) due to exaggerated centralization of production and operational difficulties connected with the most modern technologies. Giarini counters one by one the criticisms of the law of diminishing returns raised in a study financed by the Union of Swiss Banks.[2]

A striking example is the use of chemical fertilizers in agriculture. The relationship between increase in cereal production and increase in fertilizers was about 15 in the fifties, it had dropped to seven in the seventies and today is below five. The yield of cultivated plants increases very quickly with the quantity of fertilizer at the beginning, but then slows down and levels out. In the US Midwest the first use of fertilizer yielded an additional 27 Kg of hybrid maize for every kilogramme of nitrogenous fertilizer. The second use yielded 14 Kg/Kg, the third nine, the fourth four and the fifth only one extra kilogramme of maize for every kilogramme of fertilizer.

Another well-known example is that of the diminishing return of fishing technology, a highly significant example because fish supplies 18 Kg per year of high quality proteic food for every head of world population.

---

[2]   Union of Swiss Banks, *Technology and economics*, UBS, Berne, 1978.

All agricultural and zootechnic activities in general show a clear trend of diminishing technological productivity. This is linked to the fact that the technological innovations of today tend towards increased energy flow or decreased genetic variability. "It is a general economic law," writes Laura Conti,[3] "that when the energy flux that invests living systems exceeds a certain threshold, productivity decreases. It is also a general ecological law that a decrease in genetic variability destabilizes the living system."

Although in the sixties the limits of energy resources and diminishing technological returns could already be clearly discerned, Italian public and private capital made the explicit choice of increasing the mean capital investment per position of employment (an obvious premise for the creation of unemployment) and of environmental exploitation (another obvious premise for environmental degradation). Therefore it chose to favour productivity in terms of labour rather than capital or energy (see Chapter 5) as a safeguard against worker pressure: nuclei and electrons do not go on strike. This choice of waste of resources makes capital unavailable for modern experiments in decentralization. The capital is all concentrated in big, energy-consuming, high technology plants: it is a capital intensive – energy-intensive choice which necessarily leads to inflation, unemployment and the destruction of nature. In this way the present process of production consumes more and more capital, resources and energy; it causes increasing damage to the environment and to health and incorporates less and less labour in its goods.

The type of uncontrolled expansion which has been carried on since the sixties, to the detriment of the environment and the health of the workers, no longer pays. We have passed the maximum of the curve and are on the downswing. It is a myth that labour productivity can continue to increase. It is now necessary to make a rapid transition to a new model of development based on low energy and capital dependence, low potential to pollute, high employment, and low

---

[3] L. Conti, *L'uomo e il suo ambiente*, Angeli, Milan, 1980.

labour and high energy productivity. *Decentralization and smallness of scale must become the rule.* Big power stations, giant plants and the industrial metropolis have seen their day: they are dinosaurs left over from the era of rapid development. To bring them with us means to take the road towards extinction. Environmental defence also means safeguarding employment, but not all occupations. Laura Conti writes: *"The workers tend to defend the job they have rather than the job they could have."*[3]

Our basic economic concepts need to be thoroughly revised. Today the value of natural resources takes first place; 'renewability' is a precious quality. Quoting Orio Giarini: "The value which matters cannot be deduced from the cost price, but must be associated with the effective use of goods and services supplied to the user: this shifts the accent onto the duration of their use."

"Activities which produce wealth but increasingly destroy the natural heritage create a negative value or 'deduced value'. There cannot be economic development without prior and accompanying human development."

The quality of our existence depends on the quality of our territory (nature, towns, country) and the quality of the human activities which it hosts. Any attack upon this quality to the advantage of a few and the detriment of everyone else must be repulsed, even if motivated by allegedly inexorable economic necessity (the 'iron laws' of the market, competition and the like). It is to the advantage of few to build a villa in a natural reserve, to maintain a factory of poisons (dioxine, bioproteins, nuclear) even if in the latter case the wages of a few workers are being upheld. As we have seen, in the long run the destruction of natural resources and the environment pays neither in economic nor social terms.

Science and technology cannot perform miracles, nor can the laws of the market: the only truly *'iron'* laws with which our culture must come to terms are the laws of nature.

Any instance of aggression to the environment, even private, should be looked into: the rights of property do not confer the right to destroy the common heritage. A tree contributes to the climatic equilibrium of the whole planet; it is part of the ecosystem in which hundreds of other forms of life participate. To cut down a tree when it is not necessary is a form of territorial abuse by the proprietor.

It is difficult to communicate to the traditional forces of the left (parties and trade unions) the fundamental concept of the *value* of natural resources and the environment, and that the defence of jobs does not justify their destruction. It has always been thought that industrial profit was based on exploitation of labour, but *today profit is essentially based on the exploitation of future generations*. The greater standard of living of the working class in industrialised countries has been paid for by consuming the millennial resources of the planet in one to two generations often with irreversible damage to the environment.

Andrea Poggio, former editor of *La nuova ecologia*, comments[4] "For all these years we have been able to concede to worker pressure that which the increasing productivity of labour made available, drawing increasingly from Nature. This unmasks the trick of the Keynesian economy."

In other words we have pillaged from nature, defrauding the new generations. "Many of the convictions and battles of the worker movement and the left need to be reconsidered," Poggio adds.

Worker claims have been paid with Nature's goods and hence drawn from the fund of future generations. In the last 30 years, the advanced industrialised countries have perpetrated an escalation in exploitation of non-renewable natural resources. These resources have been largely depleted in an extremely short biological time, and great ecological changes (e.g. the climate) limit their further abuse. The economic and industrial system has shown its total incapacity to think beyond details and the short term. The dominating political blocks (USA and USSR) have made the same mistakes, worshiping the GNP idol and taking its increase as a measure of development and well-being. Today the GNP is no longer an index of welfare but only

---

[4]   A. Poggio, 'Crisi dell'ambiente, crisi dell'energia'. *Il manifesto*, **73**, March 29, 1980.

of cost. It may have had some validity as an economic index last century, but today we are faced with a paradox. The GNP combines in a single index values such as natural resources, capital, energy, technologies, work to provide services, the utility of these services. The only positive item is the last one; all the others should be put on the debit side of the balance. The natural and social limits of growth clearly show that the GNP has become a myth.

The definition of welfare requires other sciences and other parameters. Universal limits are posed by thermodynamics and the law of entropy on one hand, and biology and the delicate equilibrium of the biosphere on the other. All man's activities must reckon with this law and these limits.

In this sense we can speak of the supremacy of biology and the great biophysical laws of nature. This is not a reductionist concept. To the contrary, the compass of natural events and biological evolution are such that a biological approach cannot fail to be global. As we have said, the timescale is graduated in millions of years and its dimensions embrace the whole planet. This is not a vaporous ideology: the natural reality and its laws with millions of proofs, are glaringly obvious. It does not deny the complexity of relationships between men, with all their irrational, social and emotional components. It simply means knowing with certainty that if carbon dioxide continues to increase in the atmosphere then at some stage the planet will no longer support life; or that the energy to extract coal beyond a certain depth will be greater than the energy contained in the coal and then it will be *impossible* to extract any more (assuming of course that it is the last energy form left); or that at some stage, the cultivable land will not be sufficient to feed the growing population and that further jungle or forest cannot be converted to agriculture without irreparably damaging the 'green lung' of life on the Earth; or that the potential of nuclear armaments is enough to erase all trace of man from the Earth. It means knowing that nature is not going to work a miracle and that the economy must be subject to nature's laws. The only reason the major economic theories have never considered these

148

constraints is that the rates of growth have been below the danger level until recently. Today, for the first time in the history of the planet, this is no longer true. Today the economy can no longer do without the laws of entropy and biology.

Limits to growth does not mean limits to development. The confusion between growth and development has created a series of misunderstandings. The title of the first report of the Club of Rome[5] (translated into Italian as 'the limits of development') was the first source of these misunderstandings. Limiting growth means guaranteeing the survival of the human species and denying the consumer society: there is nothing reactionary about this. Limiting the growth of population, arms, energy waste, pollution, radioactivity levels etc. does not mean denying development in terms of a modern technologically advanced society (with low energy consumption), in terms of quality of life (development of quality not quantity), in terms of propagating equal distribution of both knowledge and wealth. Social justice is not to blame, but the crazy way of producing and of managing the planet of advanced capitalist societies and a certain leftist 'culture' willing to barter health and the environment for a few coins or a handful of short term jobs. This attitude is blatantly short-sighted: it does not see the repercussions on the life of future generations, the risk to which today's decisions expose our children and grandchildren. To defend jobs in the giant industrial agglomerations which pollute, in the consumeristic manufacture of trucks and steelworks, in the armaments or nuclear industry is reactionary, short-sighted and obtuse in the light of our biological understanding.

It is not necessary to renounce development and welfare, however courage is needed to completely restructure production: decentralization, soft and appropriate technologies, agriculture and handicraft. The society of tomorrow needs people who produce more food rather than more cars. Automobile production needs to establish itself at a level which maintains a constant number on the road, and public transport should be potentiated.

---

[5] D.H. Meadows and D.L. Meadows, *The limits to growth*, Universe Books, New York, 1972.

We should adopt the biological model of steady state: evolution with minimum production of entropy and maximum thermodynamic efficiency. The steady state in biology does not mean renouncing development. Natural history is the history of biological evolution, of systems in continuous dynamic variation. *The steady state does not negate evolution; it guarantees it.* If any other limits are needed, they are limits to speed. If we are social beings, let us slow down evolution as Laura Conti suggests.

From this point of view, to recognise the supremacy of biology and the limits imposed on growth by the laws of nature takes on a revolutionary quality. It introduces into economics an astounding novelty; it opens new perspectives and leads to new hopes. It enables us to attack the consumer society at a higher level based on understanding of a universal culture which has always been a property of man.

Contemporary economic theory is still linked to the positivist mechanism and Newtonian cosmology: the concepts of entropy, indeterminism and uncertainty which are the heritage of modern science, have not yet undermined the erroneous presuppositions of determinism and certainty in economic theory. Unlimited scientific progress countering diminishing returns is assumed as a postulate, and research and development are confused with miracles. Nicholas Georgescu-Roegen[6] has brought economic theory to the frontier of contemporary scientific culture. He re-examines the notion of economic value in the light also of Eugene Odum's[7] energy flows, and bestows a cosmic dimension upon the economic law of diminishing returns in the light of modern thermodynamic analysis.

In this way we see that energy from coal is a stock and solar energy is a flow. No generation can touch the solar radiation which belongs to future generations. By contrast, the availability for each generation of non-renewable terrestrial resources, is influenced by the consumption of preceding generations. A quantity of wood or

---

[6] N. Georgescu-Roegen, *Energy and economic myths*, Elmsford, Pergamon Press, New York, 1976.

[7] E.P. Odum, *Fundamentals of ecology*, Saunders, Philadelphia, 1971.

plants which grow will be available to future generations, but every Cadillac or instrument of war means fewer ploughs for some future generation, and implicitly, fewer human beings. The number of years estimated for known resources, most of which cannot be recycled, is as follows: copper 36 years, aluminium 100, iron 240, lead 26, mercury 13, tin 17, zinc 23 (Massachusetts Institute of Technology). These are small numbers on the biological scale.

The concept of value must be completely redefined. The laws of biology say that not all industries are the same and that the same economic laws cannot be used for an automobile which goes on renewable energy from biomass and one that goes on petroleum.

Marcello Cini suggests that economics could find a breath of fresh air in ecology. This might revive it from the coma in which it languishes. A scientific revolution in economics is needed.

At this historical crossroads between survival and destruction of the planet, the environment and future generations can no longer be excluded from the market. *The economy cannot continue to be based on reversible sciences (mechanics) but on sciences which penetrate into the future (biology, thermodynamics).* Living systems do not possess the determinism of technology. Living systems cannot be reduced to a quantity. *Classical economics is a form of reductionism.*

The relationship between economics and ecology has not yet been established, but certain points are clear: (a) ecology teaches economics that there are economic costs which are distant in space (the whole planet) and time (future generations); (b) ecology reveals that many environmental and human costs cannot be reduced to economic units; (c) if production only obeys classical economic laws, that which is produced will not necessarily be to man's benefit.

A fundamental key to a better understanding of the links between economics and ecology is provided by the analysis of energy flows of Howard and Elizabeth Odum. It gives a clear demonstration of the relationship between energy and money.[8] The authors note that the

---

[8] H.T. Odum and E.C. Odum, *Energy basis for man and nature*, McGraw-Hill, New York, 1981.

energy used in work is a constant measure of what has been obtained, i.e. an excellent measure of value. Much work is done by ecological systems, the atmosphere, geology. If there is less money flow, the energy flow diminishes. An increased circulation of money, by the addition of more money, raises the energy flow only if the energy resources are large. If they are scarce, or at least limited, the increase in money creates inflation. Even if money remains constant but the energy resources decrease, there is inflation.

The Odums examined many different energy systems and concluded that: (1) renewable resources of constant force generate continuous growth; (2) limited renewable resources provide growth up to a saturation point, after which a steady state is established; (3) non-renewable resources favour a period of growth which is followed by decline. The human economy can be described by a combination of types 2 and 3.

Energy flow is analysed by the Odums in a series of complex diagrams which take many parameters and functions, and their extrapolations in time, into account. For example in considering the question of production, the calculation includes not only the different types of production, but also various internal and external limiting factors, marginal factors, concentrated or dispersed forms of energy, grade of utilization of the energies etc.

W.J. Davis[9] discusses the relationships between economics and ecology on the basis of the analyses of the Odums and Commoner. He agrees with Commoner on the fundamental interactions between the three systems (natural, economic and productive) and the not coincidental occurrence of three crises at the same time (ecological, energy and economic) (See Figure 2, Chapter 1). The economic system apparently depends on the wealth created by the productive system which in turn depends on the resources supplied by the ecosystem. This is the only overall analysis which could be valid for a modern reformulation of the economy. Within its framework, the relationships between energy, resources and money must be studied. The

---

[9]   W.J. Davis, *The seventh year*, Norton, New York, 1979.

money flow goes in the opposite direction to the flow of materials, in the sense that money is exchanged for goods. Both cycles (money and resources) are carried on with the degradation of energy. The analysis of Davis reaches the same conclusions as the Odums: when the resource cycle slows down (limited resources) the result is inflation. Davis describes this process in an elegant example of a hypothetical economy in which the only goods are orchids: for the same money invested, the crisis of the resource cycle brings the price of an orchid from $1 to $2 with an inflation rate of 100 per cent.

Davis proposes a new economic science, like a garment made to measure for the ecosystem, the fundamental laws of which are those of ecology and thermodynamics. In agreement with the theories of Schumacher[10] (*Small is beautiful*), he defines the aim to be the ethical distribution of ever-diminishing resources according to the basic needs of man.

The Odums too propose a new economic model of 'steady state with constant moderate energy flow'. Personally I do not completely agree with their final solutions, just as everything Schumacher and Davis say does not convince me, but I think that some of their proposals are very interesting: (a) elimination of industries which stimulate growth and consumerism; (b) a minor role for banks; (c) miniaturisation of technologies to use less energy; (d) smaller institutions; (e) less transport, work closer to home, decrease in commuting, greater use of water and rail transport; (f) recycling of materials; (g) small decentralised electric power stations; (h) agriculture based on natural systems and mechanisms of recycling with less use of fuel.

A vision of the steady state becomes clearer. Other authors however give it different definitions and valencies. I think it is important to underline here some of the characteristics of the steady state which lead to the definition of H.E. Daly,[11] which is the one which satisfies me most.

---

[10]   E. Schumacher, *Small is beautiful: economics as if people mattered*, Harper and Row, New York, 1973.

[11]   H.E. Daly, *Steady state economics*, V.H. Freeman, San Francisco, 1977. See also Herman Daly and John B. Cobb, Jr., *For the common good*, Beacon Press, Boston, 1989.

There is nothing immobile about the word 'steady'; it does not imply poverty, return to candle-power, denial of development or scientific research. 'Steady' simply means symbiosis of man and nature. This implies the development of a system based on natural, renewable flows of energy and natural resources, without accelerating the growth and destruction of the environment and non-renewable resources. As we have seen, paths based on non-renewable resources lead to immediate but false prosperity which is inevitably followed by rapid and irreversible decline.

However these presuppositions are not enough to guarantee a livable future unless accompanied by at least two other conditions: (1) equitable distribution of wealth and world resources among peoples and persons; (2) zero population growth.

Let us return to Daly's[11] definition of steady state which embraces both Marxian and Malthusian lines of thought. It agrees with Marxists that there must be a limit to inequality and that social justice is a precondition for ecological equilibrium in all non-totalitarian societies. Birth control, without reform of property rights may reduce the number of poor but will not eliminate poverty. In agreement with Malthusians, the steady state recognises that without population control of mankind and its physical manufactures, all other social reform will be cancelled by the growing weight of absolute or Malthusian scarcity. Instead of maximizing production, a good economic aim would be a sufficient level of wealth, efficiently maintained and allocated, and equably distributed.

What to do? Control methods of production and what is produced. Scale down production and technology to the human and environmental dimension, abandoning the capitalist and socialist block economic models of continuous growth and the myth that resources are unlimited. As G. De Rita writes in the *XVII Censis Report*,[12] we no longer need any schemes, there is plenty to be done with new creativity for the future. In order to understand our position and

---

[12]   CENSIS, *XVII rapporto sulla situazione sociale del paese*, Angeli, Milan, 1983.

where we are headed, we need to marry immaterial problems of fear and values with patently material problems such as hunger and unemployment. The answer certainly is not the return to candle-power, but neither is it the specious and momentary creation of employment possibilities by building giant plants or plants involving high capital investment, such as coal and nuclear power stations, dams, ports and industrial complexes. These drain capital from the development of alternative resources and sooner or later lead to inflation, unemployment and environmental destruction rather than the survival of the human species.

Political, social and economic choices first need to be subjected to an ecological-economic filter: if an investment does not fulfill certain essential requirements indicated by Daly or Odum type analysis, it should be considered bio-economically risky or wrong, and refused. On this basis, opposition to the construction of nuclear power stations has the same social and political importance as opposition to building new bombs, unless we want to choose the path that leads to a growing throng of unemployed, an aristocracy of protected workers and another group of precarious workers who do the unsavoury jobs.

This coexistence of inflation and unemployment has led to the coining of the term 'stagflation', which indicates a situation of stagnation that nobody knows how to solve. According to Commoner, the capitalistic system, the self-proclaimed guardian of the augmenting American standard of living, can only survive, and perhaps not even then, by reducing this standard. Our leaders admit the poverty of power.[13]

Another important question which influences our approach to a different model of development is the question of time. The French green movement is well aware of this, making it the subject of a number of papers, for example by André Gorz. In *La révolution du temps choisi*[14] we read that if free time gradually exceeded time at work

---

[13]   B. Commoner, *The poverty of power*, Alfred A.Knopf, New York, 1976.

[14]   A. Gorz, *La révolution du temps choisi*, Michel, Paris, 1980. See also A. Gorz, *Adieu au prolétariat*, Edizioni Lavoro, Roma, 1982.

then our lives would be transformed. Instead of work being the aim of our existence and the carrier of dominant values, free time would become the essential time and reason for living, carrier of common values. The authors go on to propose abolishing imposed timetables, leaving this to individual choice. They add that it is necessary to abandon the mentality of compulsory productivity and pre-established working hours, giving all salary earners the choice of reducing the time they work (and the corresponding pay). The employer can only oppose this under certain conditions. Gorz asks whether this is utopian, and concludes that it is not. In Germany already 20 per cent of white collar workers freely choose their starting and finishing times. It is only a question of organisation and auto-organisation.

Georgescu-Roegen[6] also sees the passing of a considerable proportion of time in an intelligent and creative manner as an important pre-requisite for his bio-economic model. This model is outlined in eight points and several premises and general considerations. The premises concern the role that demand can play instead of supply in the realisation of the model, and the fact that man is unlikely to sacrifice the satisfaction of his natural or acquired needs, completely renouncing industrial comforts. To return to our theme at the beginning of the chapter: yes to the hi-fi, no to the chronovisor.

The most important points made by Georgescu-Roegen, in my opinion, may be summed up in these two questions: (a) Can we change our attitudes and begin to have sympathy with the people of the future? (b) Will humankind consider a programme that implies limiting its dependence on comforts and consumerism? Or perhaps another species without spiritual ambitions, the amoeba for example, will inherit the planet Earth, still awash in a sea of sunlight?

Let us now pass to the eight points of Georgescu-Roegen's programme: (1) prohibition of the production of instruments of war; (2) bringing the developing countries up to a reasonable standard of living, but not with our technological model; (3) gradually reducing population; (4) avoiding energy waste; (5) recovering from our unhealthy passion for extravagant gadgets; 6) freeing ourselves from fashion (to buy a new car every year and refurnish the house every two is a bioeconomic crime); (7) making goods more durable; (8) freeing ourselves from the 'razor syndrome'" which consists of shaving faster so as to have more time to work on a razor which shaves faster so as to have more time to work on a razor which shaves faster so as to have... etc. *ad infinitum*.

Another good example is Bertrand Russell's pin factory:[15]

> "Suppose that, at a given moment, a certain number of people are engaged in the manufacture of pins. They make as many pins as the world needs, working (say) eight hours a day. Someone makes an invention by which the same number of men can make twice as many pins as before. But the world does not need twice as many pins: pins are already so cheap that hardly any more will be bought at a lower price. In a sensible world, everybody concerned in the manufacture of pins would take to working four hours instead of eight, and everything else would go on as before. But in the actual world this would be thought demoralizing. The men still work eight hours, there are too many pins, some employers go bankrupt, and half the men previously concerned in making pins are thrown out of work. There is, in the end, just as much leisure as on the other plan, but half the men are totally idle while half are still overworked. In this way, it is insured that the unavoidable leisure shall cause misery all round instead of being a universal source of happiness. Can anything more insane be imagined?"

I personally think that it is necessary to make the transition to a low energy economic model based on renewability, appropriate technologies (which I define below), respect for the environment, population control and steady state. The era of industrial civilization

---

[15] B. Russell, *In praise of idleness and other essays*, Allen & Unwin, London, 1935.

and unlimited growth is over. We are in a phase of transition. Our choices in the next few years can make this transition less dramatic and more agreeable. Certain aspects of the transition need to be handled in a delicate manner. The main points are the transition from population increase to zero growth, a shift of employment from industry to agriculture and services, from the metropolis to villages and small towns, from specialisation to diversification at work and culturally, from centralisation to decentralisation, from consumer values to new values of life as a part of nature, from hard to soft technology, from large to small scale.

The so-called appropriate technologies can play an interesting role in this transition. They are low-capital-intensive – high-labour-intensive technologies aimed at full and correct use of the human, natural and energy resources of a given territory. They involve modest initial capital investment and use mainly renewable and decentralised energy. They are small scale and very flexible.

Biology and informatics can be protagonists in the realisation of appropriate technologies, but may also be dangerous. The negative potentialities of genetic engineering and informatics are enormous, for example they tend to increase the gap between industrialised and developing countries and can easily manipulate public opinion. The socio-political and economic-environmental consequences of their use should always be carefully evaluated.

A study of the Italian Energy Agency (ENEA) estimated the employment possibilities offered by the new technologies in Italy: 200,000 positions for technicians in the field of energy saving and renewable energy, 100,000 for environmental technicians in control and waste disposal, 200,000 experts in biotechnologies, 300,000 experts in new agricultural techniques. For informatics, the estimate is no less than 450,000 new positions.

For a new model of development to take off, a number of specific choices must be made. In Tuscany there is a saying that you cannot have the barrel full and the wife drunk. The so-called Saint-Geours report[16] of the EEC failed to find any sign of a tendency towards substituting energy with work, counter to the trend of the past two centuries.

The well-known Brandt report[17] focused on another terrible facet of the problem: (1) a modern tank costs more than a million dollars, the cost of 1000 schoolrooms; (2) a jet fighter costs 20 million dollars, the cost of establishing 40,000 pharmacies; (3) 0.5% of the world military budget would be enough to pay for all the agricultural equipment necessary to increase food production and almost reach self-sufficiency in countries having a food deficit.

Finally it is clear that the nations of the third world must seek forms of development which differ from those of the industrialised west. The Chinese model or the model of Gandhi are good examples. The comment of the Brazilian government: "Come and pollute here" should be replaced with: "Pay us royalties to conserve the Amazon rainforests."

It is impossible to export our technologies and know-how to the third world; it would only lead to wild cultural and environmental imbalances. At the most, only 20 per cent of the world population could be provided for. The myth of the interdependence of economies needs to be relinquished: the strong economies always destroy the weak. The second myth to abandon is that there is no development without growth. The third is that modern centralised industry can be controlled. Once we have got rid of these three myths, we can set out towards a new model of development, but let us not forget that the road from eco-nomy to eco-logy is not an easy road.

---

[16]   EEC, *In favour of an energy-efficient society*, Brussels, 1979.
[17]   UNO, *North-south: a survival programme*, New York, 1977.

CHAPTER 10

TO BE OR NOT TO BE:
HYPOTHESIS FOR A RENEWABLE LIFE

amlet's line can be interpreted in another way. The question for man is precisely whether or not to be. If we opt for the road to Samarra it means the end of the species Homo sapiens, and probably the end of life on Earth. This tragic modern dilemma is directly linked with the theme of Erich Fromm's book "*To have or to be?*"[1] Fromm who with Adorno and Marcuse founded the famous Institut für Sozialforschung in Frankfurt, discusses many of the questions that we have examined here. The interesting thing is that Fromm is not a biologist, physicist or ecologist, but a leading thinker in the humanistic field. In this chapter I want to show that the seeds of a new ecological view of the world have existed for a long time in many different fields.

In a collection of writings of the American Indians (Wovoka, *The revolutionary message of the natives of America*) I found this passage: "...The culture of a people is founded on the knowledge of how to survive in a specific environment. Nature as a whole is modeled by every being as the form of water is modeled by the fish, and each of our movements creates waves and transformations. Nature is an organism: it is everywhere. The westerners try to represent it by dividing it and laying it out in a line to look at it in pieces. They always seem like people outside who are trying to see what is inside. To be open to nature, to abandon oneself to it, to dissolve in it, to flow and take form with it: this is the way we create our identity without creating anything. Many people do not understand that the natural world is not a free world as the westerners understand freedom. The natural world works according to natural laws and there are many

---

[1]   E. Fromm, *To have or to be?*, Harper & Row, New York, 1976.

cycles of the natural world with which it is necessary to live in harmony. What we have to seek is freedom within these cycles and laws. This kind of freedom is difficult to imagine and it is much greater than the freedom which most people know." This is a philosophy which is not very different from that of the epigraph in the preface of this book: "Beings will never cease to be born from each other, and life is not the property of anyone, but usufruct of all". The alternative is nature kept in test tubes like the "trees of London, masterpiece of the old museum of the waters and the forests" (Jacques Prévert). However before we look at the contributions of these different cultural milieux I would like to make a few preliminary comments.

Freedom is a sort of flag for me, and my own freedom is very dear to me, but this does not stop me from recognising that in nature there are certain universal limits: for example we cannot go below -273°C, we cannot exceed the speed of light, everything that exists is in three dimensions and no more. All these limits have always existed and will continue to exist; they are limits which man cannot do anything about. To obey and respect the laws of nature does not mean renouncing one's freedom.

I am an incurable optimist but my view of the world does not allow me to draw optimism from believing that God will somehow solve the problems of the planet Earth; nor does my optimism blind me to the fact that certain catastrophes could soon put an end to man's existence and transform the planet into a wasteland. My optimism does not come from the type of scientific ignorance which induces people to say that nature is great and strong and can withstand man's onslaught. Nor is it based on a shallow hope of some technological miracle. It is based on the conviction that man has the capacity to make the right decisions which will get him out of this dilemma.

I do not agree with the imposition of sacrifices or denial of 'needs'. This however does not mean favouring the future construction of the *chronovisor*, or passively accepting the present society, planned in the name of productive rationalism against the laws of thermodynamics and biology. Industry needs to make choices rather than defend the

162

status quo to the bitter end. Social behaviour also means respecting the life of others and of the environment; means postponing immediate economic interests in favour of the survival of our children. It means having the social courage to ask for the closing down of factories which pollute, of power stations which emit radioactivity and of the armaments centres.

I am also convinced that any reductionist operation is very dangerous and that man, unlike animals, has a socio-cultural history. This however does not mean that he has to interpret things in a solely socio-economic framework or become entrenched in a type of humanism which refutes the basis of science. In the first case there would be danger of an even more dangerous reductionism: *economic reductionism*; and in the second case, *humanistic reductionism*. "Be scientifically literate," wrote C.P. Snow[2] exhorting his readers to bridge the gap between scientific and humanistic thought. It is time for new alliances, as Prigogine calls them, alliances between the history of man and the history of nature, between historical and biological time scales, between Marx and Malthus. It is time for a new culture and for new needs. The near future needs scientific humanism and ecological culture. This is the only hypothesis suitable for the species Homo sapiens. It is non-reductionist, a hypothesis for a life based on quality, equilibrium with the environment, the renewability of resources: in a word, for a 'renewable' life.

In the introduction to *To have or to be?* Fromm points out that economic progress has remained a property of rich countries, and that the gap between rich and poor countries is wider than ever. He also points out that technical progress has brought ecological dangers and the risk of nuclear conflict. Either of these could put an end to our whole civilization and perhaps to all of life. Fromm continues that the belief that unlimited technological and industrial progress would bring happiness to everyone by satisfying all desires, and would re-establish social peace and the harmony of man with nature, has now been proved false. Contemporary man has become a cog in the

---

[2]   C.P. Snow, *The two cultures*, Cambridge University Press, Cambridge, 1964.

immense bureaucratic machine, he is alienated and manipulated by industry, the mass media, governments; he is exposed to ecological dangers and the risk of nuclear war; he is psychologically depressed, isolated, anxiety-ridden, prey to destructive impulses. Fromm continues that technology has revealed its other face, the face of the goddess of destruction (like Kalì of the Indus). The major powers continue to build nuclear weapons of greater and greater destructive capacity, without succeeding in coming around to the only sensible solution. All nuclear weapons and the atomic power stations which supply material for nuclear warheads should be dismantled. Practically nothing is being done to avert an ecological catastrophe. No concrete measures are being taken towards the survival of the human race.

The rift between the mechanism of the technological and productive processes and man as a natural being is at the root of many of the things that are wrong today. Fromm notes, however, that this rift between disciplines did not exist in key figures like Albert Einstein, Niels Bohr, Werner Heisenberg and Erwin Schrödinger who were deeply interested in philosophical and spiritual questions.

Starting from the presupposition that to have a lot is not the same as living well, Fromm gives a series of indications for the construction of a new society: (a) avoid industrial centralization; (b) give up the goal of unlimited growth; (c) give up the free market economy; (d) promote scientific progress but ensure that its practical applications are not dangerous for mankind; (e) restore the possibilities for individual initiative; (f) substitute psychic satisfactions for material gain; (g) separate scientific research from its industrial and military applications.

The last point means that the goal is human welfare. This rarely coincides with profit and never with military aims. If the social climate changed, Fromm adds, it is certain that the transition from egoism to altruism would not be so difficult. Fromm speaks of 'humanistic religiousness', a type of religiousness that has been in preparation from Buddha to Marx. It is non-theistic, non-institutionalized, not counter to socialism or existing religions. He speaks of a genuine humanistic socialism, a synthesis between the spiritual nucleus of the late medieval and the development of the rational thought of science.

But prior to Fromm there were already two great ecological works: *Brave new world* and *1984*.

While the youth of the twenties dreamed with Verne about the domination of nature and the conquests of technology, Julian Huxley (grandson of the biologist Thomas Henry Huxley, apostle of Darwin in England) was studying the relationship between resources and population in biology, and his brother Aldous Huxley was writing *Brave new world*.[3] Since Huxley nobody dreams about Captain Nemo, but wide open eyes are focussing on the void which the technological society is opening before us. The novels of Verne do not appeal to the imagination of today's children. The film *Blade runner* becomes our ally, along with the marvelous science fiction comics of the Latin American authors.

On a different level, the false note sounded by Newton, Bacon and Descartes, and the excesses of Comte in separating thought from material reality, man from nature, today need to be replaced by what Waddington[4] called 'biological sagacity', or Sartre 'the biological origin of man'.

According to Rifkin,[5] every time a businessman, politician or scientist speaks in public today about some pressing problem, it is as if his words were written by these thinkers of the past (Bacon, Descartes, Locke, Smith). Their opinion was that the faster nature were transformed, the greater our progress, the more ordered the world, the more time saved. Today the law of entropy tells us that the converse is true.

Descartes made mathematics the most powerful instrument of knowledge. Rifkin goes on to say that the mathematical world of

---

[3] A. Huxley, *Brave new world*, Chatto & Windus, London, 1932, and *Brave new world revisited*, Chatto & Windus, London, 1959.

[4] C.H. Waddington, *Man-made future: problems of the twenty-first century*, Saint Martin's Press, New York, 1978.

[5] J. Rifkin, *Entropy – a new world view*, Viking Press, New York, 1980.

165

Descartes was colourless, odourless and tasteless. It could not exude, drip or overflow. Mathematics represented total order. In the world of Descartes, everything had its place and all relationships were harmonious. It was the world of precision, not confusion. The mechanical age was characterised by this concept of progress. When reduced to its simplest abstraction, progress is seen as the process by which the disorderly natural world is harnessed by man to create a more ordered material environment.

It is this mechanistic concept of the world, the Cartesian and technological concept, which is leading us and the planet to extinction. This does not mean refuting science as many humanists do. Without science there is not even a probability of understanding man and the world. The complexity of possible choices demands the fusion of the two cultures on a higher plane, it demands scientific humanism.

The separation of natural laws into different sciences is a human artifice: nature is a whole. As we have seen, uncertainty, irreversibility, complexity and instability reign in nature. "Instability", writes Giuliana Conforto,[6] referring to Prigogine, "permits the creativity and free choice of the system." To simplify man, to reduce him to Homo faber in production time, a mere instrument of productivity, means destroying him and leading him to destruction, because it eliminates his capacity to play, to ponder, to create.

As Rifkin puts it, the need for profound cultural change is felt everywhere by those who do not identify with this culture of death in which there is no time; our high energy culture has so compartmentalized our minds that we find that we are no longer in harmony with nature, the source of life. Some type of scientific humanism is required like that so present in Aldous Huxley and D.H. Lawrence.

In 1966 *Brave new world* was still prohibited reading in some US high schools. This provides a measure of the importance of Huxley's work. Far from being utopian, it is a penetrating satire of a dehumanized society, planned in the name of rationalistic production.

---

[6]   G. Conforto, 'Lo stato stazionario, l'evoluzione e il tempo'. In *Energia, ambiente e transformazioni sociali*, Edizioni Grafiche Palermo, Rome, 1982.

It is the first testimonial of that interest for the fate of man to which the writer eventually dedicated himself completely.

In the thirties, Huxley had already sensed the dangers of genetic engineering and described a process which resembles cloning. The horrendous new era described in *Brave new world* is characterised by the productive-organisational revolution symbolized by the name of Henry Ford. People are divided into castes: the alpha plus class consists of technicians; the attractive Lenina, obviously with reference to Lenin, is one of them. In this 'new world', the individuals of the species are created and raised in the laboratory. Anxiety and depression no longer exist: 'opium is the religion of the people' (Marx's famous saying inverted). The happiness of *Brave new world* is the happiness of hedonism administrated by the accountants of civil living, also known as sociologists. The hedonist is always an accountant of happiness, unlike the passionate who is a gambler.

A quarter of a century later, Huxley wrote *Brave new world revisited*,[3] outlining his sociopolitical convictions. Today, some of his comments are remarkably modern and provocative. Let us examine some of them: (a) the impersonal forces of overpopulation and superorganisation, the social engineers which these forces seek to direct, are leading us towards a new medieval system; (b) democracy can hardly be expected to flourish in a society in which economic power is increasingly concentrated and centralized; (c) in the course of evolution, nature has seen to it that every individual is different. Any culture which tries to standardize the human being in the interests of efficiency or in the name of any religious or political dogma, commits an offence against the biological nature of man. Every individual is biologically unique. Hence freedom is a great good, tolerance a great virtue and regimentation a great calamity; (d) ...establish and develop a world policy for the conservation of land and forests; create suitable surrogates for our fuels, if possible less dangerous and less depletable than uranium...; (e) anyone who does not want the spiritual impoverishment of individuals and society should leave the metropolis and revive the small provincial community.

Huxley also wrote that the pill had not yet been invented. He asked himself what the eventual reply would be to the objections of the Roman Catholic Church. He observed that time was against us in the race between natural resources and population increase.

Huxley and D.H. Lawrence were friends for many years. They often met during their stays in Italy, in Florence and Rome. Not long before his death, Lawrence was a guest of the Huxleys at Forte dei Marmi. Huxley was to edit the first collection of Lawrence's letters.

The author of *Lady Chatterley's lover* identifies the industrial society as the last trick of the 'white subject' to complete his plans of absolute dominion over things. He compares him with the 'love subject' who is no longer subject because he is liberated by love from his subjectivity; he is attuned to the vibrations of the trees and waters, flowers and clouds. In *The plumed serpent*[7] and *The woman who rode away*[8] Lawrence reveals the subjugation and gradual extinguishing of nature, the desire of the 'white subject' to have dominion over matter. In the introduction to *The woman who rode away*, we read that this means "separation, colonization, dominion, abstraction, mechanization, necrosis, negation of the body, of ecstasy, of pleasure, of joy of the flesh and finally the death of nature". For Lawrence this death arises from the meeting of the Christian idea of the supremacy of inner conscience with bourgeois and industrial rationalism which has gradually gained a hold through Descartes, Hegel and positivistic scientism.

Rivers of ink have been written about Orwell's *1984*[9] with the pretext of finding or negating a similarity between the world of today and that described in the famous novel of 1948 (84 was merely an inversion of 48). Orwell was of Scottish extraction and his father a British official in Bengal. He fought in the Spanish civil war with a small Catalan anarcho-syndicalist group and he had a strong aversion for the Soviet regime and Stalinism. Some of his visions were truly prophetic: the division of the world into areas of influence, wars

---

[7]   D.H. Lawrence, *The plumed serpent*, Martin Secker, London, 1930.

[8]   D.H. Lawrence, *The woman who rode away*, B.Tauchnitz Leipzig, 1929.

[9]   G. Orwell, *1984*, Harcourt Brace Jovanovich, San Diego, 1984.

which break out every day in different places, reciprocal invitations to moderation exchanged by the super-powers, closed circuit television monitors, slogans and prefabricated phrases. One of many examples: in *1984* war is peace, freedom is slavery, ignorance is strength. Begin and Sharon called their aggression of Lebanon 'peace in Galilea'; the USSR called its armed intervention 'support for the Afghan people'; Reagan called his nuclear warhead missiles 'pacifiers'. The nuclear-telematic apparatus and the ambiguous surveillance that informatics exercises are two of the Orwellian aspects of our society.

ECOLOGY OF MIND

In his book *Steps to an ecology of mind*,[10] Gregory Bateson claims that the underlying causes of the present wave of environmental perturbations lie in the combination of: (a) technical progress; (b) population increase; (c) traditional but erroneous ideas about the nature of man and his relationship to the environment. On the nature of man, I consider that another energy source which is fast running out is man's psychological capacity to resist aggression by a world governed by exclusively economic and technological laws. On one hand there is the "drama of man trying to control, with an equivalent inner force, the pressures of the technological and industrial world which he created and which have grown out of all proportion".[11] On the other hand there is loss of 'I', collective anonymity to which man is condemned by the lack of differentiated stimuli resulting from mass production and consumerism.

Thus while the system requires more and more strength, it distributes weakness. This is another equation which does not balance and leads to the dehumanization of man, his inner collapse. Even the excessive specialisation required by our system increases the danger of this. Man belongs to an ecological cycle and also to a social environment, both extremely complex.

---

[10]   G. Bateson, *Steps to an ecology of mind*, Chandler, San Francisco, 1972.
[11]   A. Nin, *Early diaries of Anais Nin*, Harcourt Brace Jovanovich, San Diego, 1980.

In order to have awareness and understanding of our part as active elements in these two worlds, the mind must receive a great variety of impressions and information, and keep the problems in perspective. Excessive specialisation confines man to his own small field. The sense of security he feels is shattered the minute he ventures abroad. In his lack of experience he is left to choose between irrational primordial behaviour and excessive rationalism. Specialisation lessens the chance that man's primordial thousand amoebic arms may meet, touch and communicate.

Science and humanities on their own cannot achieve what is required. We need more than interdisciplinary collaboration, more than the integration of different fields: new disciplines and new channels of human communication are necessary if man is to attain 'happiness', not in Christian terms (happiness not of this Earth) nor in hedonistic terms, but in terms of equilibrium between man and his environment, with his own kind and with himself. This is the fraternity indicated by Sartre's appeal: "Sub-men, do not despair!" and the need for science and the humanities to recognise a common matrix.

I give this curious example to illustrate this idea of coming together. The famous dancer and symbol of art itself, Isadora Duncan, once at the railroad station met Ernst Haeckel, the German biologist who in 1866 coined the term 'ecology'. An immediate rapport sprang up between them. For Isadora, everything about him emanated a subtle perfume of health, strength and intelligence, if one can speak of perfume of intelligence.[12] Haeckel likened Isadora's dancing to all the universal truths. Two different worlds found a way to communicate.

Today many writers and artists, whether aware of it or not, are pervaded by ecological sensibility. See for example the novels of Amado and Garcìa Marquez.

In Italy there was Italo Calvino, with his profoundly naturalistic *Barone Rampante*.[13] As we have seen, Calvino understood the differences between the visions of Monod and Prigogine. While underlining the

---

[12] I. Duncan, *My life*, Boni & Liveright, New York, 1927.
[13] I. Calvino, *Il Barone Rampante*, Einaudi, Turin, 1982.

importance of science, it is necessary to avoid positions of the 'medieval technocratic fideist' like Monod who was convinced that nuclear energy was indispensable for human survival, and that nature was destroyed because there was not enough technology rather than too much. Man does not "stand at the edge of the universe in which he has to live, like a gypsy: a universe deaf to his music, indifferent to his hopes, suffering and crimes" (Monod). His only solution is not faith in an aseptic and omnipotent science; to the contrary, "scientific knowledge purged of its reveries of being an inspired supernatural revelation, may today be found to be a poetic sensing of nature and at the same time the natural process of nature, an open process of production and invention in an open, productive and inventive world" (Prigogine). A dialogue between man and nature: notice that Prigogine refers back to philosophers like Lucretius, Leibnitz, Bergson who examined this dialogue in depth. It is a dialogue between three systems, states Gregory Bateson: the organism of each man, human society and the vaster ecosystem. The point is the role of awareness in combining these three systems which seem to be under different laws. Bateson draws from Lewis Carroll: by imperfect pairing of biological systems, Alice is partnered by a flamingo and the ball by a porcupine in a game of croquet. Alice does not understand the 'flamingo system' and vice versa, both mistake fireflies for lanterns, as the saying goes. The problem of linking man with his natural environment or with society by awareness, is similar.

It is not a small problem. What changes in cultural coordinates are necessary? What social agents will be able to manage the transition to a society based on profoundly different laws and values? More than ever we need to define, create, utilise a new form of knowledge to build a new model.

The coming generation will certainly play a leading role, and so will the ecological movements. 'Think globally, act locally.' 'Not to the right or to the left, straight ahead.' 'Love thy future as thyself.' These are the slogans of the green movement. With the last slogan, G. Baget-Bozzo emphasises that our reference should be 'the unknown child of the third millennium'. If we make future generations our problem, if the species decides to plan its survival, then an alliance between man and nature is required, a new culture founded on ethical-cultural grounds, not material growth, consumerism and productivity.

THE SUPREMACY OF BIOLOGY

We have seen that the risks we run today have planetary dimensions: climate, radioactivity, nuclear war, population increase, food shortage. We have also seen that nearly all these problems involve the future generations and imply decisions which will affect our children and grandchildren.

These two aspects, the planetary dimension and the long term, do not fit into the classical political framework. Our historical-political experience does not cover problems of this type: we are caught unprepared.

We need a big cultural operation, a synergic effect of expertise and political and cultural assets. The foundations cannot be other than a closer study of the biological aspects of natural equilibria, the evolution of man and of behaviour. When I speak of the supremacy of biology I do not mean that policy should be guided by an aseptic science, but that policy be permeated and nourished by biology.[14]

We have seen that there has been a real increase in 'welfare' in western industrial and, partially, in the eastern socialist countries in the last fifty years. Although in these countries (which constitute only a small proportion of the world population) the living conditions of the working class have in fact improved, this has been paid for in three negative ways: (a) the rapid development of technology has increased the gap between rich and poor countries, privileged and depressed areas: our welfare is based on an increasing death rate from hunger and an increasing number of unemployed and marginalised people;

---

[14] *"The rules of law are accessory, those of nature essential; those of law are agreed upon, not native, those of nature native, not agreed upon."* (Antiphone the Sophist, 5th century BC).

(b) increased organisation or 'technological order' (see Chapter 2) has inevitably created more disorder in the environment which means irreversible damage to the planet; (c) worker claims have been conceded by increasing the demand on nature's non-renewable resources; in other words *exploitation of the 'future' class*, the generations of our sons and grandsons. The effect of this is to expedite the approaching economic and resources crisis which could be very serious indeed. These crises could be avoided by rational use of resources and a transition from a system based on non-renewable to one based on renewable resources.

We are on the verge of a great transformation, which the scarcity of resources makes inevitable. If we begin to lay the foundations now, the transition will be smoother, less dramatic, with less likelihood of catastrophes like war, famine and disease.

We are the children of the industrial revolution, which brought considerable advantages to mankind, especially at first. Saturation of benefits was soon reached and now the balance weighs heavily on the negative side.

What we need now is a cultural revolution. This will be generated by the scarcity of natural and energy resources and our children will be the protagonists. The sense of this revolution will be to overturn many of the values which our present society considers inviolable: (a) the concept of renewability: any human or technological act based on the renewability of matter and energy is ethically valid; vice versa any act based on non-renewable resources is an error towards our children, and could even be regarded as exploitation of them; (b) 'to be' must replace 'to have' as a basic social value and for the satisfaction of our needs; quality of life must replace quantity; (c) the laws of thermodynamics must become our guide in production decisions, even in relation to economic processes; (d) we must acquire the concept of 'limited growth' and biophysical equilibrium (or steady state) as an obvious consequence of the fact that we live on the planet Earth; (e) to contribute to population increase must be considered ethically reprehensible (no more than two children per couple). This is a blueprint for a new and renewable way of living.

Ubiquitous renewable biological resources constitute the foundation of this 'renewable lifestyle'. Because they are distributed throughout the world they offer the potential of equipartition between peoples; they favour self-fuelled systems rather than energivorous; they favour the maximization of choice rather than the suppression of diversity; they favour the spread of individual responsibility rather than bureaucratic or hierarchical centralization; they favour interdisciplinarity and the survival of many cultures rather than specialisation and monoculture.

Inside this new model of living, the greatest possible liberty and diversity of behaviour must be stimulated within an overall organisation of economics and resources, so that the mass concept can be replaced by personalized roles and values. André Gorz[15] writes that ecological imperatives (the depletion of resources) make necessary an economic, social and cultural revolution which abolishes the constraints of capitalism and establishes a new relationship between man and the collectivity, the environment which surrounds them, nature. We are not painting nature in golden terms or 'returning to nature', but realising the simple fact that man's activity comes up against an external limit in nature.

Gorz continues that without a struggle for different technologies, the struggle for a different society is in vain. "The choice of nuclear energy, in a capitalist or socialist system, implies a centralized, hierarchical police state. Socialism is not immunized against technofascism. Socialism is not better than capitalism if it uses the same tools: the domination of nature by man inevitably leads to the domination of man by technology."

André Gorz adds that capitalistic technological development has progressively bent the worker to the requirements of production, rather than freeing him. The modern worker has even lost the participation in the productive process that the worker of yesterday enjoyed by virtue of his technical ability.

Today work is an activity extraneous to the worker, and in which

---

[15] A. Gorz, *Adieux au prolétariat au-delà du socialisme*, Editions Galilée, Paris, 1980.

the development of the individual is no longer possible. Gorz proposes a 'utopia'. With the whole system of production to revise, what we need are not old ideas but audacious new ones ('utopia or death') to stimulate discussion. He thinks that it is impossible to restore interest and creativity to work like that of the production line, but speaks of the possibility of creating an economy in two sectors, progressively reducing the market production sector which *per se* necessitates repetitive alienating work (perhaps even with the help of computers) and simultaneously expanding the field of autonomous activities, in which individuals can produce goods for their utility value. Utopia? Perhaps.

Or perhaps the utopians are those who are still convinced that an increase in production can improve the quality of life and that such an increase is materially possible. None of these people seems to realise that beyond the limits of material growth there are 'social limits to growth'.[16] Social scarcity is evident for goods and services which tend to deteriorate the wider their use: if everyone stands on his toes, nobody gets a better view. When material growth causes an increase in the number of cars, there are traffic jams and the commodity 'car' is no longer a 'good'. "Classical industrialists can by now be regarded as dreamers," Gorz declares. "They must be to talk of growth when the price of energy and metals has begun to multiply; when the lack of fresh water means that seawater has to be distilled; when we learn that half the marine fauna and flora filmed by Jacques Yves Cousteau in 1956 no longer existed in 1964; when the wood and paper industries are destroying the Amazon rainforest with the blessing of the Brazilian technocrats, thus striking at its source the very thing that regenerates 25 per cent of our atmospheric oxygen. Material growth has physical limits. Every attempt to exorcise them by recycling or treatment, only postpones the problem because operations of regeneration also require energy. Ecological realism breaks with economic rationalism." The less utopian industrialists are moving their industries and pollution to poor countries. A report of

---

[16] F. Hirsch, *Social limits to growth*, Harvard University Press, Cambridge, Mass., 1978.

the Rand Corporation discloses that by the end of the century, the US will have all of its manufactured goods made abroad, keeping only service industries and research at home.

Much of the new model of development (not growth) is still unconceived. As Kuhn remarked, the passing of time often uncovers anomalies that the existing theory can no longer explain. The divergence between theory and reality can become very great, causing serious problems. This is exactly what is happening today between existing socioeconomic theories and the natural reality of the planet. The only way out is a new paradigm, a revolutionary theory: cultural and theoretical commitment becomes a necessity together with a role for new intellectual currents and disciplines other than economics. The project, based on new theoretical horizons, new conceptual tools, new scientific advances, must operate taking into account the natural constraints. The project is difficult but not impossible to carry out. It is a project in which the process of the liberation of man and his individual happiness will play a fundamental part. It is a wager for the man of the millennium 2000. It is a challenge to be won if we are not to meet the fate described in this closing story.

'FLIGHT CODE'

Taken from a comic strip of the Latin American designer Alfonso Font, 'Flight code' is symbolically dramatic. It contains all the contradictions, knots and paradoxes of our society, demonstrating that science and technology can improve life or destroy it, depending on how they are used.

A space ship leaves with two people on board. It is loaded with a hundred tons of manam seeds, rich in calories, which are to feed the 40 million starving inhabitants of Central City for a year. Forty million lives depend on the success of the mission. At a certain stage, a meteorite strikes and punctures the external oxygen pipes and the oxygen leaks away in a myriad of bubbles into space. After the initial shock, the two men aboard attempt the impossible, transforming the space ship into a hydroponic culture of manam seeds in every possible container, in order to create oxygen. They calmly provoke bradycardia, breathe more slowly and give up smoking, but despite all possible expedients they realise that they will have enough oxygen for nine days, whereas it will take fifteen to reach Earth.

There is a solution, but a terrible one. One of the two must be sacrificed so that the other can reach Central City and save the |inhabitants. They cast lots and one of the pilots bravely lets himself be sucked out into space via the rubbish chute. He dies in a few seconds and his body begins a macabre orbit around the space ship, like a satellite. The other pilot, after interminable days, reaches Central City at the end of his oxygen and the limits of his forces, and requests 'absolute precedence over absolute precedences' to land. Ground control requests his flight code, to which the desperate pilot replies that he hasn't one, but must land with maximum urgency. He is told not to worry, there is no urgency, the monitor finds nothing out of order on the space ship; however identification, the flight code, is necessary.

The exasperated pilot shouts: "The flight code is on my co-pilot who has been dead for fifteen days. He is orbiting this space ship and I see him pass the port hole every hour. He sacrificed himself so that this cargo could reach Earth to save 40 million people from starvation. I have only a few minutes' oxygen left, forget the procedure, let me LAND!" The reply from the control tower is: "We are sorry, we need the flight code."

"Who is there? Who am I speaking to? I am without oxygen. Who are you? Tell me your name!" shouts the dying pilot.

"I am Robot Controller RC-9 of the spaceport, at your orders!" A robot! The desperate pilot makes the only possible decision: he leaves standby orbit and begins to land. Ground control immediately notifies infraction of the procedure and the order is given: "Intercept! Four, three, two, one, zero ... strike!" The missile hits the space ship and destroys it in space.

End of procedure.

CHAPTER 11

HISTORICAL TEMPOS, BIOLOGICAL TEMPOS REVISITED

AGAINST LAWS

Today the clock of hours
hangs around my neck on a cord;
The course of the stars and sun,
the cock's crow, shadows,
cease today,
and everything that announced time
is mute and deaf and blind:
all nature becomes silent
to the tic-toc of law and hour.
(Friedrich Nietzsche)

Twenty years have gone by since I began to write *Historical tempos, biological tempos*.

While science celebrates unimagined achievements at the turn of the millennium, it is clear to all that we are living a crisis that manifests in many forms, especially as a lowering of the quality of life, including psychological quality, as the destruction of nature and as an increase in unemployment among the young, despite economic and technological growth.

On rereading and rewriting *Historical tempos, biological tempos* again after 20 years, I find that the problems discussed are still topical, and are if anything emphasised by subsequent events and the subsequent evolution of political and scientific thought.

The revisited text, with this final chapter on the challenges of the new millennium, is also an opportunity to look more closely at the revolutions of scientific and philosophical thought in this last period, from Ilya Prigogine to Gregory Bateson, and from Herman Daly to Stephen Jay Gould.

BIRTH OF THE EARTH

4,600 MY	Jan 1	Birthday
3,500 MY	April 1	$CO_2 + H_2O \rightarrow CH_2O + O_2$
500 MY	Oct 30	Ozone
400 MY	Nov 8	21% O_2
350 MY	Nov 15	Soil Formation
200 MY	Dec 26	Dinosaur
65 MY	Dec 30	Dinosaur death, Meteoritic Bombardment
1.7 MY	Dec 31, noon	Hominidae Appearance
0.0002 MY	Dec 31	The Industrial Revolution

I owe this table to a Japanese scientist; it is a birthday greeting to our Mother Earth, after 4,600,000,000 years (MY= million years before present).

Time is the key for a rediscovery of laws and history, evolution and myths, biodiversity and beauty.

Rereading the book I wrote twenty years ago spurred me to write this new chapter, in which I look at how various aspects and disciplines interweave: sustainable economics and evolutionary physics, ethical values and environmental policy, aesthetics and the science of complexity. It also gave me the pleasant opportunity to turn again to Bateson, Daly and Prigogine, who have long woven fabric of this kind. It also spurs me to state, with a mixture of embarrassment and pride, that the forecasts of 20 years ago have turned out to be well-founded and scientifically correct.

Today, after 300 million years of stability, a drastic change in the carbon dioxide content of the atmosphere is observed: millennial balances between feedback and homeostasis that guaranteed the two cycles on which life is based, the carbon and oxygen cycles, have been upset. Carbon dioxide concentrations have continued to increase as predicted and have passed 370 ppm. Global warming is causing enormous masses of water to evaporate and giving rise to extreme events, typically out of season.

E.C. Lorenzini of the Harvard Smithsonian Center for Astrophysics, Cambridge, Massachusetts (USA) made the following predictions based, among other things, on data from the latest 'tethered' satellites: a future increase in mean global surface temperature between 1°C and 4°C before the year 2100 and a doubling of atmospheric carbon dioxide concentrations in the same period. The most probable mean value is therefore an increase of around 2.5°C in about 100 years. To understand what an increase of this size means, Lorenzini invites us to consider the climatic history of the Earth. Palaeoclimatic studies based on air trapped for thousands of years deep in the ice of Antarctica and Greenland show that the mean temperature of the Earth increased about 10°C in the 4000-year period from the end of the last ice age, which occurred 14,000 years ago. The largest change in the mean temperature of the Earth, measured over a period of the order of a century, since the last ice age, was therefore about 0.25°C per century. The change predicted for the coming century, due to the increase in the greenhouse effect precipitated by man, is therefore 10 times greater than historical values.

Living species adapted to the slow increase in temperature but it is reasonable to suppose a ten-fold rate of change will be less easy for life to adapt to.

Chlorofluorocarbons dispersed in the atmosphere tear the ozone layer which has been stable for millions of years. Car exhausts and the emissions of thermoelectric power stations have increased the acidity of rain twenty-fold. Every minute, 40 hectares of forest disappear at the hand of man, a loss of 15 million hectares of greenery per year. At this rate, the tropical forests will disappear within the span of a human life, with serious effects on atmospheric and climatic equilibria.

The population continues to grow: 300 million persons at the time of Christ, 600 million in 1500, doubled again in 1800; two and a half billion people in 1950, doubled again in only 40 years to the present figure of over six billion. The planet is severely stressed in the briefest of biological times. How many forests, leaves, green areas are necessary to support the life of 10 billion people? For the first time in the history of man, the conditions which enabled life to emerge and evolve on earth are threatened.

I recall a seminar held by myself and father Ernesto Balducci at the citadel in Assisi. A young man objected to what I had to say about population increase, pointing out that the Bible says to increase and multiply. Ernesto brushed my arm to indicate that he would answer: "Read carefully", he said. "It says, *Grow, multiply and fill the Earth! Today, the planet is full!*"

This was one of the forecasts of 20 years ago: six billion people in 2000. A fundamental consideration arises here, namely that *infinite growth (of population, consumption, production) cannot exist on a finite planet*. Let us look at this idea and the concept of 'unsustainable growth'.

The scientific novelty is that the system in which we live, the planet Earth, is a finite system, and as such has constraints: land constraints, constraints regarding absorption of wastes and pollutants, constraints related to the great cycles of life (air, water, oxygen and so forth), constraints that limit indiscriminate increase in population and production. Physical reality is therefore subject to constraints which determine limits. For example, if the population increases, more food is needed. To have more food, a greater yield per hectare of land is necessary, which impoverishes the soil and causes erosion, contamination of aquifers and eutrophication of seas. Alternatively, forests can be cleared to create more land for food production, but this means loss of biodiversity, stress on the carbon and oxygen cycles (green-house effect) and climatic change, which in turn affect agriculture.

Again we find underlined the difference between the rapid pace of technological growth and the slower tempos of the biological sphere, at the core of the environmental crisis.

Twenty years ago I used the term 'carrying capacity' coined by Herman Daly with reference to the planet Earth. This term refers to the capacity of the planet to sustain a growing population. In the years that followed, with a series of encounters with Odum, Costanza and others in different parts of the world, the idea of 'generational solidarity' began to take root, and the verb 'to carry' was replaced with 'to sustain' (implying duration in time) which gave rise to the idea of sustainable development.[1] In French, the adjective 'durable' was adopted.

SCIENTIFIC NEW WORLD REVISITED

Millions of years have elapsed since our great grandmother, the bluegreen algae (as Laura Conti called it), brought about the photosynthetic revolution, giving rise to life on the Earth. If we look at this great story of evolution, from which we sprang, we discover three protagonists ignored by the dominant physical sciences: beauty, time and biodiversity.

The limits of a purely quantitative view of nature, which denies the fundamental ecological category of quality and the importance of the aesthetic element, are becoming evident in the face of the complex temporal dynamics of the biosphere and global ecosystem. These temporal dynamics are based on multiple relations in a process of coevolution based on forms, colours, sounds, odours and flavours. The history of nature is a systemic and evolutionary history, one in which quantity and quality are ever present, a story in which the aesthetic element plays a determinant role.

---

[1] E. Tiezzi and N. Marchettini, *Che cos'è lo sviluppo sostenibile?*, Donzelli, Rome, 1999.

A science of nature cannot ignore all this. A modern evolutionary physics will have greater need of Mandelbrot fractal geometry than of Euclidean geometry; it will have more need of the thermodynamics of Prigogine dissipative structures than of Einstein relativity.

Two interesting new aspects of fractal geometry are the global (generalist, not specialist, says Mandelbrot) approach and the fact that one begins with nature and constructs a geometry of nature (of ribs, clouds, crystals, galaxies etc.) upon it, using geometric objects previously regarded as 'esoteric', mere curiosities or mathematical pathologies. The geometry of nature is chaotic. Fractals enable us to see something more in the chaos of the biosphere, using stochastic processes and thus combining chance with choices, exactly as happens in the great process of biological evolution.

Observation of nature tells us two important things: that quality and time are not external values but intrinsic properties of living organisms. This is the great lesson of Darwin's theory of evolution, a theory which, among other things, has the merit of not indicating aims or certainties for evolution. Darwin repeatedly underlined the fundamental role of chance and the absence of an end towards which life as a whole moves.

Time modulates forms and structures, sounds and colours: these properties are an uninterrupted thread throughout evolution. Colours, forms and structures pass information between different living species, between plants and animals, between us and the environment. 'Objective science' has reduced colours to a mere parameter, blue to radiation with a wavelength of 440 nm, just as 'subjective art' regards colour as subjective, as mere sensation.

If we want to see an orange also as a blue orange, without giving up the orange, if we want to contaminate science with art and art with science, we need to consider relations, connections and interactions between observer and observed. The colour green of a plant exists irrespective of whether we see it, because it existed for millions of years before the human mind. Before man appeared on the Earth it was recognised by millions of plants and animals. However, it is also true that when we look at a plant, our mind begins a series of interpretations, of rational and intuitive synthesis, and weaves a series of relations with the plant itself and with the emotions engendered by it. All this is influenced by our culture and our genes.

What we have to do is to fuse microscopic and macroscopic, supersede the dichotomies of reductionism and antireductionism, study biological phenomena in terms of relations and self-organisation, so that the behaviour of the parts becomes coherent. We need a philosophy of nature that I would define as Lucretian, in which rediscovery of aesthetics is determinant in science, economics and politics. It means weaving the first new alphabets that enable us to converse with nature.

As urged by Bateson, it means demolishing the anti-aesthetic assumption which arose from the importance initially attributed by Bacon, Locke and Newton to the physical sciences, the assumption that all phenomena can and must be evaluated only quantitatively. Quality and form also have scientific value.

Quality and time have played a fundamental role in biological evolution, contributing to the evolutionary success of species and modelling the forms of life. Today, these two categories, taken as the foundation for an evolutionary epistemology, are real values of systems ecology to consider in scientific education and for decisions regarding sustainable development.

The study of living systems shows the determinant role of time in the transformation of structures (both molecular and biological) and the role of form in relations between species. In this way the value 'quality' is recovered, underlining the basic scientific contribution of aesthetics in nature.

In a new non-linear, systemic culture, weave and narrative add clarity and complexity to scientific thought. Time, understood as number of relations experienced and as information stored as energy-matter, models the molecular forms of biological evolution.

Time is not an abstraction, it is an integral part of matter; it is part of what exists and not a single political, social or economic theory can be described without considering the irreversibility of time. The problem of time is so fundamental, because the structures that connect us to the rest and which are an integral part of the coevolution of nature and the human mind, embody time. Without time it is impossible to explain how these structures change and proceed. It is not a question of renouncing rationality of thought, but rather of recovering a series of ethical and aesthetic values, of getting to the bottom of what Bateson called ecology of mind, and of asking ourselves about the structures that connect sequoias and crabs, crabs and sea anemones, anemones and amoebas, amoebas and schizophrenics, and all this with us. In other words, we should ask ourselves about the structure that connects man and nature in this complex story of biological evolution. We should reconstruct the values and ways of thinking that lead towards integration of man and nature, bearing in mind that we are simultaneously 100 per cent nature and 100 per cent culture. The foundations which will enable us to accept the challenge of complexity and ecology can only emerge from such awareness.

Taking a coevolutionary viewpoint, environments, niches and species evolved with different time scales and different rhythms, in the framework of the great laws of nature that existed aeons before the mind and ideological models of humans. Nature is never the same: it changes, and in changing, is continuously sending information to the mind of man, and man, through his decisions, converses with and modifies nature: two entities, man and nature, in continual exchange.

The universe is made of relations between matter and energy. Music, sounds, words are energies that weave relationships between different biological species: in this splendid exchange, the aesthetic component is essential. To reduce sound waves to mathematical models and quantitative measurements is to lose most of biological reality, to the detriment of science and understanding. Aesthetics means overcoming a purely quantitative approach in science and introducing the basic ecological category of quality. Aesthetics is important for quality of life. The much hoped for change in direction of civilization must be founded, among other things, on aesthetic values.

Western science has put nature in a cage of geometric rules and mechanistic laws. We know that they do not hold for living systems, ecosystems, events of biology and events of ecology.

In the two great cultural revolutions of physics this century, quantum theory and relativity, the two limiting characteristics of Newtonian laws, determinism and reversibility, were included. Neither irreversibility nor the creative role of time were accorded scientific dignity. This has caused a schizophrenic dualism between being and evolving, between the static description of the physical world and the irreversible behaviour of the living world.

To aim for an evolutionary view of the Earth also means going in the direction of unification of the two cultures (science and humanities). Science has given too much space to space, and has ignored time. In history, in human affairs, in ecology and so forth, the role of time is fundamental: *memories are certainly more important than kilometres.*

Again it was Prigogine[2] who wrote:

"The classical view of science led to a dichotomy: in 1663 Robert Hooke promulgated the statute of the Royal Society, defining its aims as being to improve knowledge of natural things and all the useful arts, manufactures, machines and inventions by means of experiments, without meddling in theology, metaphysics, morality, politics, grammar, rhetoric or logic. In this, already we observe the famous division of the 'two cultures' made famous by C.P. Snow.[3] The viewpoint with nature re-emerging today will hopefully overcome this opposition between interest in nature and interest in man."

The new evolutionary physics breaks away from the safety of determinism and/or subjectivism and adds irreversibility and

---

[2]   I. Prigogine in E. Tiezzi, *The essence of time*, Wit press, Southampton, 2002.
[3]   C.P. Snow, *The two cultures*, Cambridge University Press, Cambridge, 1964.

uncertainty to its basic paradigms. In other words, it accepts the stochastic nature of time as an intrinsic property of matter. The view common to classical and quantum mechanics, based on the reversibility of time, is a simplification. It is now evident that nature has instabilities and chaos: physics can no longer ignore it.

If we succeed in bringing evolutionary physics and aesthetics together, science and art, ecology of complex systems and philosophy, we can begin to be creative, which is a value indispensable for research in any field.

The aim of science should be to live in harmony with nature, but sometimes the scientist (or rather, the sorcerer's apprentice who uses the results of science for profit, control and power) has made discoveries that could damage man and the environment.

An example for the turn of the millennium is the attempt to clone human beings. Clonation or replication of living species (plant or animal) is the opposite to biological evolution and the origin of life, both of which are based on biodiversity and the diversification of forms, individuals and biological species.

The risk associated with genetic engineering (future mutations, environmental impact, destruction of other species) is unknown and has yet to be studied. The effects on future generations and on the planet are unpredictable. A plant engineered to defend itself from viruses and insects (very ecological, according to the sorcerer's apprentices, because we can save on insecticides) could also be a killer of useful microorganisms and other living species.

The seeds of transgenic plants often require activators which are patented and held by biotech multinationals, who could hold farmers and consumers to ransom by limiting their supply. Other seeds contain the so-called *terminator* gene which makes seeds of the next generation sterile (as in Monsanto transgenic maize). The farmer therefore has to buy seed every year from the multinational, which gains control of world food production.

A double market seems likely: on one hand American, Swiss and Japanese multinationals (Monsanto, Novartis etc.) that will offer transgenic food, initially at low cost (to feed the world, they say, but the third world will actually be their gigantic guinea pig); on the other hand, hopefully, the market of Europe defending its agriculture, its food products, its culture linked to the land and culinary traditions by centuries of history.

Ethically, as Hans Jonas pointed out, genetic engineering is an agent to modify pre-existing structures, the reality and genus of which are the primary datum; reality and genus are neither invented or produced *de novo*, but the fact of discovering them enables them to be perfected by new inventions. Thus we have partial rather than total production, modification of the project rather than a project, and the result is not a finished artefact but only a tiny fragment.

There is also the question of reversibility and irreversibility. In mechanics, everything is reversible. Living organisms, on the other hand, are characterised by the irreversibility of their modifications.

Traditional engineering can always correct its mistakes, and not only during design and testing: even the finished product, for example cars, can be sent back to the factory for correction. This is not true for biological engineering. Its acts are irrevocable. When its results manifest, it is too late to do anything. What is done, is done.

Jonas concludes by asking what will be done with the inevitable accidents of genetic manipulation? This ethical problem needs to be tackled and solved before a single step is taken in that direction.

We are told that man has always modified nature genetically, but it is usually omitted that this was done selecting desired qualities in plants or animals through their *phenotype*, never by directly manipulating their *genotype*.

The great difference in 'engineering' the genes of living organisms lies in the fact that the blueprint of a plant or animal is within it, is unique for every individual and is passed down from generation to generation. The blueprint of a machine is the same for all machines of that type and it stays on the engineer's desk.

189

PHILOSOPHICAL NEW WORLD REVISITED

The voice that spoke was certainly that of our master.
He knows how to bring together traces dispersed here and there.
... He observed the stars and traced their position and orbits in the sand;
he watched the sea of air, never tiring of its clarity, movements clouds, lights.
... He considered men and animals, seated on the shore, he looked for shells.
... He linked distant things. Now the stars were men, now men were stars,
stones were animals, clouds were plants; he played with forces and phenomena;
he knew how and where to find and evoke them.
(Novalis, Disciples of Sais)

The illuminist dream of reason that dominates nature, rather than living in harmony with it, has generated the monsters of *one-way, levelling thought (pensiero unico)*, a type of thought that does not observe the times and modes of nature, that does not know its constraints and limits. Yet it is from limits and constraints that creativity springs, artistic and scientific creativity. Freedom is not of this world; it is not part of our nature. Nature is made of spatial and temporal limits and constraints: our life is not eternal, our dimensions are three, our body weight is what it is, likewise our possibilities of movement. We could say that the beauty and diversity of evolutionary history lie in the fact that every living species has different limits and constraints. Some do not walk erect, others only move in water, others live in air ... Biodiversity consists in the fact that every plant, animal and human has different constraints and has to learn to live with them. They are life itself, what determine diversity, without which art and science would not exist because creativity comes from being subject to constraints.

Some delude themselves that the three famous values of the French revolution (liberté, égalité, fraternité) are absolute values. Freedom, however, is limited by the biophysical constraints of the planet, equality by respect for the fabulous diversity of the planet, and fraternity cannot be limited to humans or to the present age, but

190

should be extended to all forms of life, plant and animal, and to future generations. Only thus will we truly be part of the history of nature.

Today we are beginning to understand the value of relations. Bateson taught us to get away from reductionist mechanistic views and look at things in terms of relations, shifting attention from subject and object (anthropocentric subjectivity) to relations, histories, relations in time. It is a cultural revolution, a scientific revolution, a change in paradigm, or rather a getting away from paradigms and the dogmatic thinking associated with them.

Ecology, a systemic and global science, proposes getting away from man/nature dualism, avoiding anthropocentric and nature-centred positions. In this, Bateson's set of intuitions was perhaps the deepest, understood not only as the study of exchanges of energy and matter, but also exchanges of information. In order to go in the direction of ecology, both culturally and socially, it is necessary to avoid absolutist positions and fundamentalist stereotypes, which manifest in the following examples, nearly always 'black and white':

(a) the myth of the objectivity and neutrality of science, together with the certainties and 'positivity' claimed to be derived from it;

(b) faith in linear conclusions of rational approaches (and the resulting antibiological dogma of excluding instinct, aesthetics, quality and emotions);

(c) hotly defended subjective individualist anthropocentric positions;

(d) denial of the reality of natural limits, claiming them to be subjective (all in the mind), as if the 3,000,000,000 years of biological evolution and networks of information predating man did not exist;

(e) naturalism taken to extremes with exaltation of the human body, human ego and uniquely anthropocentric point of view;

(f) deification of nature and faith in its power to repair all damage inflicted, which unfortunately is not true (which is why green movements exist).

These are substantially absolute positions which do not admit different points of view. Ideologically, they deny or deify either natural reality or the role of the subject, they dispense certainties (though the *complexity* and *uncertainty* of natural events make certainties unlikely), or deny the existence of reality (though the biophysical *limits* of ecology tell us that natural reality and the observer must both be considered).

Indeed man is inextricably bound to the ancient fabric of nature: his learning processes are part of nature's complex and largely unmapped history; the time the human species has existed on the planet is a mere flash compared to the dimensions of biological time.

History is giving ecology the chance of being protagonist of a change of paradigm, a very special type of change, because it does not presume or envisage construction of a new paradigm, but simply escape from the rigidity of old Newtonian-Cartesian assumptions and the anthropocentric view, both lacking the parameter time and the interaction man-nature-society.

Ecology, as a privileged observatory required for the survival of humans on the planet, can take up the environmental challenge, the role of policy-making and philosophy-making. It can take the dimension of a great cultural change in relations between humans and nature, overcoming rationalism, mechanicism, illuminism, to build a science of complexity, uncertainty and constraints, based on evolutionary-epistemology and evolutionary-physics. An ecological science, between evolution and conservation, is in line with the poignant comment of Renzo Piano:

> "By working, you grow, and quite soon you learn that the
> words modernity, progress and growth are infernal traps,
> in the name of which we are continuously defrauded."

ECONOMIC NEW WORLD REVISITED

To grow means 'to increase naturally in size by the addition of material through assimilation or accretion'. *To develop* means 'to expand or realise the potentialities of; bring gradually to a full, greater or better state'. In short, growth is quantitative increase in physical scale, while development is qualitative improvement or unfolding of potentialities.

In Italy, especially in economics schools, the two terms are regularly exchanged and confused. Emblematic of this is the famous error: the book of the Club of Rome *Limits to growth* was translated into Italian as '*limits of development*'.

What does this mean for the environment and the biophysical resources of our planet? First of all, *infinite growth cannot exist on a finite planet.* Secondly, as Rudolph Clausius, the father of thermodynamics, pointed out, in the economy of a nation there is a general law: *in a given period, do not consume more than is produced.*

The integration of economics, ecology and thermodynamics gives rise to the new theory of ecological economics, or sustainability, which has nothing to do with that bad oxymoron 'sustainable growth'.

The theory of sustainable development arose as a radical and frontal counterposition to the theory of globalisation. The two golden rules on which it is based are due to Herman Daly:[4]

(1) *harvest rates should be equal to regeneration rates (sustained yield);*

(2) *waste emission rates should be equal to the natural assimilative capacities of the ecosystems into which the wastes are emitted.*

Regenerative and assimilative capacities must be treated as *natural capital*, and failure to maintain these capacities must be treated as capital consumption and therefore not sustainable.

The natural resource flow and the natural capital stock that generates it are the *material cause* of production; the capital stock that transforms raw material inputs into product outputs is the *efficient cause* of production. One cannot substitute efficient cause for material cause: one cannot build the same wooden house with half the timber, no matter how many saws and hammers one tries to substitute.

According to Daly[4], certain biases

"have prevented us from seeing the obvious, namely that
the fish catch is limited by remaining populations of fish
in the sea, not by the number of fishing boats; timber

---

[4] H.E. Daly, *Steady state economics*, V.H.Freeman, San Francisco, 1977. see also Herman Daly and John B. Cobb, Jr., *For the common good*, Beacon Press, Boston, 1989.

production is limited by remaining forests, not sawmills. More sawmills and fishing boats do not result in more cut timber and more caught fish. For that you need more forests and larger fish populations in the sea. Natural capital and man-made capital are complementary and natural capital has become the limiting factor."

Today we are in a transition phase between an 'empty-world' economy and a 'full-world' economy; the only path to sustainability involves investing in the scarcest resource or limiting factor. Sustainable development means investing in natural capital and scientific research into the global biogeochemical cycles on which sustainability of the biosphere depends.

Daly writes of three communities: between people, with non-human species and with future generations. He criticises growth-orientated economics which is culminating in environmental crisis, and lays the foundations for a new economics and a new social ethic. The basis of this ethic is community with the future (or generational solidarity). We want to leave our grandchildren a planet that can still sustain community life. This leads to the basic concept of sustainability, sustainable lifestyle, sustainable development.

Sustainability refers to the set of relations between human activities with their dynamics and the biosphere with its generally slower dynamics. These relations should be such that human life can continue, individuals can satisfy their needs and the various human cultures can develop, but keeping man-made changes within limits so that the global biophysical context is not destroyed.

Calculations on natural capital (*Nature*, **387**, pp. 253-60, 1997) indicate that the wealth produced by humans from nature is worth about $US 33,000,000,000,000 per year, compared with a gross global product of $US 18,000,000,000,000 per year. The ridiculous thing about orthodox economic theory is the persisting belief that a growth of three per cent (from 18 trillion to 18.6 trillion dollars) would solve world economic problems, when natural capital has been eroded from about 50 to 33 trillion dollars. This is why, despite the widely proclaimed growth of the GNP, the gap between industrialised and developing countries has widened. It is the reason for the growing pockets of poverty in rich countries. It is the main reason for the increase in unemployment among young people.

Daly writes:

"Overpopulation is thought to be cured by the demographic transition. When GNP per capita reaches a certain level, children become too expensive in terms of other goods forgone and the birth rate automatically falls. Economic growth is the best contraceptive, as the slogan goes. Whether the product of increased per capita consumption times the decreased birth rate of 'capitas' results in increasing total consumption beyond optimal scale remains an unasked question. More concretely, is it necessary for Indian per capita consumption to rise to the Swedish level, for Indian fertility to fall to the Swedish level, and if so, what happens to the Indian ecosystem as a result of that level of total consumption?"

"Unjust distribution of wealth between classes, we are told, is rendered tolerable by growth, the rising tide that lifts all boats, to recall another slogan. Yet growth has in fact increased inequality both within and among nations. To make matters worse, even the metaphor is wrong, since a rising tide in one part of the world implies an ebbing tide somewhere else.'

Daly concludes:

"The end of physical growth, or even growth of a value-weighted index of physical growth like GNP, is not the end of progress... "

"Growth in GNP can in fact be uneconomic."

This 'full-world' growth pushes each country to further exploit the remaining global commons, and to try to grow into the ecological space and markets of other countries. *This collective folly we call* '*globalisation*', Daly says.

Twenty years ago, *Historical tempos, biological tempos* tried to bridge the gap between scientific knowledge of the great biological cycles and economic policy, between evolutionary physics and new ethical values. Some seeds have germinated, in a stochastic manner, as is usual in the history of biological evolution. A few seeds may have responded too vigorously; others fell on stony ground.

I would like to conclude indicating some good reading matter to future generations: Laura Conti, Vandana Shiva, Ernesto Balducci and Rigoberta Menchu. My best wishes for a long life also go to 'the people of Seattle'.

NAME INDEX

The Essence of Time

E. TIEZZI, University of Siena, Italy

With the ever-accelerating pace of daily life, the modern age seems to demand that science, too, should respond at speed. In this book Tiezzi highlights the continuity between the physical-mathematic and humanistic sciences. He also urges us to reflect on the tempo of the modern era, and to contrast it with the brilliance and complexity of the human relationship with a living world.

A guide to the key scientific ideas of our time which relate ecology with economy with the laws of thermodynamics, and those that illuminate an understanding of the human relationship with planet earth, the text traces themes such as entropy to negentropy, flux of energy and material and information. The result is an exploration of great scientific depth and the most complete historical survey to date of the ideas behind ecological economics.

Series: The Sustainable World, Vol 2
ISBN: 1-85312-949-6 2002 apx
200pp
apx £59.00/US$89.00/€94.00

WIT*Press*
Ashurst Lodge, Ashurst, Southampton, SO40 7AA, UK.
Tel: 44 (0)23 8029 3223
Fax: 44 (0)23 8029 2853
E-Mail: witpress@witpress.com

Sustainable Energy

Y. PYKH and I.G. MALKINA-PYKH, Russian Academy of Sciences, Russia

Partial Contents: Energy Technology; Energy Planning and Policy; Energy and Society; Energy and the Global Environment; Future Global Aspects; Glossary of Terms.
Series: The Sustainable World, Vol 3
ISBN: 1-85312-939-9 2002 apx
150pp
apx £57.00/US$88.00/€92.00

Sustainable Water Resources

Y. PYKH and I.G. MALKINA-PYKH, Russian Academy of Sciences, Russia

This book, and its two companion volumes (see below), provide general guides for those in planning, administration, or other disciplines who require an overall view of the subjects involved.
Contents: History and Introduction; Systems Analysis of Water Systems; Natural Water Resources; Water Technology; Water Economics; Water and Society; Water and the Global Environment; Water and the Future; The Method of Response Function as a Modelling Tool; Glossary of Water Terms.
Series: The Sustainable World, Vol 5
ISBN: 1-85312-938-0 2002 apx
130pp
apx £49.00/US$75.00/€78.00

Multifunctional Landscapes

Volume I - Multifunctional Theory, Values and History

Editors: J. BRANDT, University of Roskilde, Denmark, and H. VEJRE, The Agricultural University, Copenhagen, Denmark

During the post-war period, intensified land use has been furthered primarily by spatial segregation of functions. Growing land pressure and environmental problems have made this strategy problematic and a paradigm of integrated multifunctional use of landscapes is emerging. This challenges a variety of disciplines and requires interdisciplinary cooperation on complex landscape research.

This book and its two companion volumes (see next column) present a collection of papers discussing these challenges from a variety of perspectives. All of the contributions also form the basis for a set of recommendations for future research within the three themes examined.

Volume 1 - Multifunctional Theory, Values and History focusses on future demands on the landscape concept, values and assessment of multifunctional landscapes, and ecological aspects of multifunctional landscapes in historical perspective.

Series: Advances in Ecological Sciences, Vol 14
ISBN: 1-85312-930-5   2002   apx 250pp
apx £85.00/US$131.00/€137.00

Multifunctional Landscapes

Volume II - Diversity and Management

Editors: J. BRANDT, University of Roskilde, Denmark, and H. VEJRE, The Agricultural University, Copenhagen, Denmark

This volume highlights monitoring multifunctional terrestrial landscapes, biodiversity versus landscape diversity in multifunctional landscapes, and complexity of landscape management.

Series: Advances in Ecological Sciences, Vol 15
ISBN: 1-85312-934-8   2002   apx 225pp
apx £75.00/US$116.00/€121.00

Multifunctional Landscapes

Volume III - Continuity and Change

Editors: Ü. MANDER, University of Tartu, Estonia, and M. ANTROP, University of Gent, Belgium

A collection of papers discussing multifunctional landscape challenges from the perspective of continuity and change. Future recommendations for landscape planning and socio-economic programmes are included.

Series: Advances in Ecological Sciences, Vol 16
ISBN: 1-85312-935-6   2002   apx 400pp
apx £132.00/US$204.00/€214.00
SET ISBN: 1-85312-936-4   apx £259.00/US$399.00/€418.00 (Over 10% saving)